Life Support:
Conserving Biological Diversity

John C. Ryan

Vicki Elkin, Research Assistant

Ed Ayres, Editor

Worldwatch Paper 108
April 1992

Sections of this paper may be reproduced in magazines and newspapers with acknowledgment to the Worldwatch Institute. The views expressed are those of the author and do not necessarily represent those of the Worldwatch Institute and its directors, officers, or staff, or of funding organizations.

Printed on 100% recycled paper containing 15% post-consumer waste

Table of Contents

Introduction

No one knows, even to the nearest order of magnitude, how many life forms share the planet with humanity: roughly 1.4 million species have been identified, but scientists now believe the total number is closer to 10 million, and it may be as high as 80 million. The majority of species are small animals, such as insects and mollusks, living in little-explored environments like the tropical forest canopy or the ocean floor. But nature retains its mystery in familiar places as well. Even a handful of soil from the eastern United States is likely to contain many species unknown to science.[1]

Biological diversity—the ecosystems, species, and genes that together constitute the living world—is complex beyond our understanding, and valuable beyond our ability to measure. But it is clear that this diversity is collapsing at rates that can only be described as mind-boggling. Difficult as it is to accept, mass extinction has already begun, and the world is irrevocably committed to many further losses. Harvard University biologist Edward O. Wilson estimates that in tropical rain forests alone, roughly 50,000 species per year—nearly 140 each day—are either extinguished or condemned to eventual extinction by the destruction of their habitat. These figures are predictions based on theory rather than tallies of individual extinctions, but as measures of the erosion of life worldwide, they may be considered conservative. Habitat losses in other areas, such as tropical dry forests and rangelands, and other mechanisms of extinction, such as over-exploitation, pollution, and the introduction of exotic species, surely push the numbers higher.[2]

Biological impoverishment is occurring all over the globe, and nowhere can it be taken lightly. Natural communities with fewer species than rain forests have, such as oceanic islands and freshwater lakes, are probably

I would like to thank Janis Alcorn, Dexter Hinckley, Reed Noss, Sandra Postel, Walter Reid, Ted Smith, Michael Wells, and Edward Wolf for their comments on drafts of this paper.

losing even greater proportions of their varied life forms. Entire ecosystems and genetic varieties within species (including both wildlife and domesticated crops) are also disappearing, likely at rates greater than the extinction of species themselves.[3]

Such collapses are not new to the planet; in fact, extinction is the eventual fate of all species. But since complex organisms first arose some 600 million years ago, individual species have lived an average of 1 to 10 million years, and the natural "background" rate of extinction has been on the order of one to ten species per year. Now, the activities of one species are multiplying that rate several *thousand* times, causing an impoverishment of life unmatched since the decline of the dinosaurs some 65 million years ago.[4]

Why should disappearing beetles, plants, or birds concern us? To biologists, and to many others, the question hardly needs asking: a species is the unique and irreplaceable product of millions of years of evolution, a thing of value for scientific study, for its beauty, and for itself. For many people, however, a more compelling reason to conserve biological diversity is likely to be pure self-interest: like every species, ours is intimately dependent on others for its well-being.

Time after time, creatures thought useless or harmful are found to play crucial roles in natural systems. Predators driven to extinction no longer keep populations of potential pests in check; earthworms or termites killed by pesticides no longer aerate soils; mangroves cut for firewood no longer protect coastlines from the erosive force of the sea. Diversity is of fundamental importance to all ecosystems and all economies.

In any meaningful strategy to safeguard the world's biological heritage, the top priority will be protection of wildlands. True protection of these ecosystems alone will require sweeping changes in the way humanity views and uses the natural world, and a commitment to limit the amount of the earth's bounty that society appropriates to itself. But in order to stave off biological poverty, humanity will have to learn not only to save diversity in remote corners of the world, but to maintain and restore it in the forests and waters that we use, and in the villages and cities where we live.

Few would argue that every beetle, every remaining patch of natural vegetation, or every traditional strain of rice is crucial to planetary welfare. But the dismantling, piece by piece, of global life-support systems carries grave risks. No one has pleaded biodiversity's case better than American wildlife biologist Aldo Leopold did nearly a half-century ago: "If the biota, in the course of aeons, has built something we like but do not understand, then who but a fool would discard seemingly useless parts? To keep every cog and wheel is the first precaution of intelligent tinkering."[5]

Life on the Brink

Biodiversity is commonly analyzed at three levels: the variety of ecosystems within which organisms live and evolve, the variety of species, and the genetic variety within those species themselves. The degradation of whole ecosystems, such as forests, wetlands, and coastal waters, is in itself a major loss of biodiversity and the single most important factor behind the current mass extinction of species.

Tropical rain forest ecosystems, which are believed to shelter at least half the planet's life forms, have been reduced by nearly half their original area. Deforestation annually claims 17 million hectares of the wet and dry tropics, an area about four times the size of Denmark. In Benin, Côte d'Ivoire, western Ecuador, El Salvador, Ghana, Haiti, Nigeria, and Togo, forests have all but disappeared. In most nations, forests have been progressively divided into small fragments surrounded by degraded land, with their ability to sustain viable populations of wildlife and vital ecological processes impaired.[6]

Brazil has more tropical forest—and probably more species—than any other nation. (See Table 1.) Massive deforestation continues there, but has slowed appreciably since its 1987 peak, thanks to unusually rainy weather, changes in government policy, and a slowdown in the Brazilian economy. With roughly 90 percent of its groves still standing, by national or international standards the Brazilian Amazon remains relatively intact. In contrast, however, the country's once-vast Atlantic coastal rain forests and the coniferous Araucaria forests of the South have been dev-

Table 1. Deforestation in the Megadiversity[1] Countries, 1980s

Country	Share of World's Land Area	Share of World's Flowering Plant Species[2]	Annual Deforestation Rate[3]	
	(percent)		(square kilometers)	(percent)
Brazil	6.3	22	13,820[4]	0.4
Colombia	0.8	18	6,000	1.3
China	7.0	11	n.a.	n.a.
Mexico	1.4	10	7,000	1.5
Australia	5.7	9	n.a.	n.a.
Indonesia	1.4	8	10,000	0.9
Peru	1.0	8	2,700	0.4
Malaysia	0.2	6	3,100	1.5
Ecuador	0.2	6	3,400	2.4
India	2.2	6	10,000	2.7
Zaire	1.7	4	4,000	0.4
Madagascar	0.4	4	1,500	1.5

[1]Twelve countries identified as harboring exceptional numbers of species.
[2]Based on total of 250,000 known species; because plant ranges overlap between countries, figures cannot be added.
[3]Closed forests only.
[4]1990 figure.

Sources: Jeffrey A. McNeely et al., *Conserving the World's Biological Diversity* (Gland, Switzerland, and Washington, D.C.: International Union for Conservation of Nature and Natural Resources et al., 1989); World Resources Institute, *World Resources 1990-91* (New York: Oxford University Press, 1990); Ricardo Bonalume, "Amazonia: Deforestation Rate Is Falling," *Nature*, April 4, 1991; Walter V. Reid, "How Many Species Will There Be?" in T. Whitmore and J. Sayer, eds., *Tropical Deforestation and Species Extinction* (London: Chapman and Hall, 1992).

astated—more than 95 percent destroyed by logging and urban expansion.[7]

Outside the tropics, a number of ecosystem types have been all but eliminated from the planet, including the tall-grass prairies of North America, the great cedar groves of Lebanon, and the old-growth hardwood forests of Europe and North America. Temperate rain forests, less widespread than their tropical counterparts, are probably the more endangered ecosystem. Of the 31 million hectares once found on earth, 56 percent have been logged or cleared. In the contiguous United States, less than 10 percent of old-growth rain forests survive, scattered in small fragments throughout the Pacific Northwest. In the rain forests of British Columbia, only one of 25 large coastal watersheds has wholly escaped logging.[8]

Wetlands, like forests, are important repositories of biological diversity both near and far from the equator. Among the world's most productive ecosystems, they help regulate water flows, remove sediments and pollutants, and provide essential habitat for waterfowl, fish, and numerous other species. They are threatened in many parts of the world by drainage for agriculture or urban expansion, conversion to aquaculture ponds, overgrazing, and, in forested wetlands, logging.

Damage to wetlands has been severe in industrial nations, with losses in Italy, New Zealand, and California exceeding 90 percent. Canada contains one fourth of the world's wetlands, and overall it has lost relatively few. But even here, major losses have occurred: Atlantic salt marshes, prairie wetlands, and Pacific estuarine marshes have all been reduced to a third of their original extent. Vast areas of bog and marsh remain in the country's sparsely populated northern regions.[9]

Mangrove swamps, which grow along about a quarter of all tropical coastlines, have suffered heavy losses in Asia, Latin America, and west Africa. In Ecuador, for example, nearly half of these protective wetlands have been cleared, mostly for shrimp ponds, and half of the remainder are targeted for clearing. India, Pakistan, and Thailand have all lost at least three fourths of their mangroves. Indonesia seems determined to follow suit: in Kalimantan, its largest province, 95 percent of all mangroves are to be cleared for pulpwood production, even though the fisheries nursed by Indonesian mangroves earn roughly seven times as much in export revenue as all their wood and charcoal production combined.[10]

Mangroves and other coastal wetlands form part of an interdependent complex of coastal habitats, protecting those inland from the erosive force of the sea and those offshore from land-based pollution. Coral reefs, among the most complex and species-rich ecosystems on earth, easily withstand the pounding of ocean waves but are highly sensitive to changes in nutrients, water temperature, and light levels. When fertilizers, sewage, or eroded soil pollute the clear tropical waters where they thrive, these communities of slow-growing animals are often killed off, smothered by sediment or overgrown by fast-spreading algae.[11]

Direct monitoring of underwater communities is difficult, but it is likely that coral reefs are in worse condition than either forests or wetlands. The most recent global survey, based on data from the early eighties, found problems such as sedimentation, water pollution, and direct damage from fishers and tourists degrading reefs off 90 of 109 countries. A decade ago, the Philippine government estimated that 71 percent of that nation's reefs—the most diverse in the world—were in "poor to fair" condition at best; only 6 percent were judged "excellent." Coastal development and deforestation in the tropics have increased dramatically in the past decade, undoubtedly burdening reefs and other coastal habitats with greater sedimentation and pollution.[12]

Corals are also exhibiting a new kind of degradation: massive bleaching. When subjected to extreme stress, they jettison the colorful algae they live in symbiosis with, exposing the white skeleton of dead coral beneath a single layer of clear living tissue. If the stress persists, the coral dies. Now considered the worst threat to the survival of coral reefs, bleaching occurred without warning at sites throughout the tropics in 1980, 1983, 1987, and 1990. The most serious cases were found in the corals of the Caribbean Sea. While its causes are not known, bleaching appears when ocean temperatures are abnormally high, leading some scientists to call it a harbinger of global warming.[13]

The second and most familiar category of biodiversity loss is the decline of species. (See Table 2.) The majority of species (and of extinctions) are invertebrates of the tropical forest, too numerous to identify, let alone monitor. For islands and other habitats with relatively limited numbers of species, the situation is somewhat easier to track. In Hawaii, for example, 41 species of Hawaiian tree snail were listed as endangered by the

Table 2. Selected Animal Species in Jeopardy, Early 1990s

Species Type	Observation
Amphibians[1]	Worldwide decline observed in recent years. Wetland drainage and invading species have extinguished nearly half New Zealand's unique frog fauna. Biologists cite European demand for frogs' legs as a cause of the rapid nationwide decline of India's two most common bull-frogs.
Birds	Three fourths of the world's bird species are declining in population or threatened with extinction.[2]
Fish	One third of North America's freshwater fish are rare[3], threatened, or endangered; one third of U.S. coastal fish have declined in population since 1975. Introduction of the Nile perch has helped drive half of the 400 species of Lake Victoria, Africa's largest lake, to or near extinction.
Invertebrates	On the order of 100 species lost to deforestation each day. Western Germany reports one fourth of its 40,000 known invertebrates to be threatened. Roughly half the fresh-water snails of the southeastern United States are extinct or nearly so.
Mammals	Almost half of Australia's surviving mammals are threatened with extinction. France, western Germany, the Netherlands, and Portugal all report more than 40 percent of their mammals as threatened. All cetaceans (whales and dolphins) are treated by the Convention on International Trade in Endangered Species as threatened or likely to become so.
Carnivores	Virtually all species of wild cats and most bears are declining seriously in numbers.
Primates[4]	116 of the world's roughly 200 species are threatened with extinction.
Reptiles	Of the world's 270 turtle species, 42 percent are rare or threatened with extinction.

[1]Class that includes frogs, toads, and salamanders.
[2]Definitions of "threatened" and "endangered" vary, but generally "endangered" means in imminent danger of extinction, and "threatened" includes species imperiled to a lesser degree.
[3]Many species are naturally rare; others have been made rare due to human activities. Rare species are vulnerable to endangerment in either case.
[4]Order that includes monkeys, lemurs, and humans.

Source: Worldwatch Institute, based on sources documented in endnote 14.

U.S. government in 1981, but today only two remain in substantial numbers, and they are declining rapidly.[14]

Little attention has been paid to conservation of such humble creatures, but their extinction—or even their removal from part of their range—can have profound consequences. Populations of American oysters, which were once so numerous in the Chesapeake Bay that they could filter all the water in the bay every three days, have fallen by 99 percent since 1870. Now, it takes a year for oysters to filter the same amount of water, one reason the bay is increasingly muddied and oxygen-poor.[15]

The Chesapeake Bay example makes clear that extinction is only the most extreme form of biodiversity loss. Natural systems, whether on land or in the water, are more than collections of species or genes: they are functioning wholes, processes as well as parts. The American oyster is not considered an endangered species, but its role in the Chesapeake demonstrates that a mere reduction in the local population of a species can disrupt the functions it performs in its ecosystem. The benefits of that system to humanity, such as clean water or seafood harvests, may thus be lost long before extinction becomes a threat.

Reflecting the widespread degradation of aquatic habitats, freshwater fish are declining in many areas. In the main rivers and great seas (the Black, Caspian, Aral, and Azov) of the southern republics of the former Soviet Union, more than 90 percent of major commercial fish species have been killed off. Similarly, in peninsular Malaysia, a recent four-year inventory found fewer than 50 percent of the 266 fish species known to have inhabited the region's rivers before the advent of large-scale logging.[16]

Unfortunately, for many aquatic life forms, the only measure available to gauge their well-being is how many of them are consumed by humans. This imperfect measure does indicate the widespread over-fishing of many coastal and open-ocean species. Catches of Atlantic cod and herring, Southern African pilchard, Pacific Ocean perch, king crab, and Peruvian anchovies have all declined over the past two decades, according to the U.N. Environment Programme. Namibian fisheries are on the brink of collapse: in mid-1991 the fisheries department said that a fishing moratorium of at least five years was needed to allow anchovies, mackerel, and other species to recover from "potentially disastrous" levels.[17]

Because most modern fishing techniques are unselective, they often cause greater damage to non-target species than to the ones sought for consumption. Drift nets, the notorious "walls of death" that stretch up to 60 kilometers in length, are scheduled to be phased out of use by the end of the year, but other fishing technologies are comparably destructive. For example, the world's shrimp trawls (funnel-shaped nets that are towed behind boats) land nearly 2 million tons of shrimp each year but also snag 2.5 to 10 times that weight in other ocean organisms. Off the northeastern coast of the United States (ironically the most ardent supporter of the drift net ban), boats trawling for yellowtail flounder catch and discard three times as much fish as they keep.[18]

Scientists have discovered an apparent worldwide decline in amphibians (species such as frogs and salamanders) in recent years. Because amphibians divide their time between land and water habitats and their skin is permeable to airborne gases, they are especially sensitive indicators of environmental degradation. Habitat conversion and air pollution are likely causes of many species' decline, but amphibians are mysteriously diminishing in seemingly pristine nature reserves. Again, the services provided by nature often become apparent only when they are lost. An adult frog can eat its own weight in insects daily, and diminishing frog populations in India have been linked to higher rates of pest damage to crops in Maharashtra state and of malaria in West Bengal.[19]

Patterns of plant diversity parallel those of animals, with two thirds of the world's plant species found in the tropics. Although prehistorical extinction spasms tended to claim mostly animals, plants too are now threatened with extinction on a large scale. Peter Raven, director of the Missouri Botanical Garden, has estimated that one fourth of all tropical plants are likely to be wiped out in the next 30 years.[20]

Outside the tropics, the greatest concentration of threatened plants is found in the arid landscapes of southern Africa. Four-fifths of the plants there are endemic (found nowhere else), and 13 percent of these—more than 2,300 species—are reportedly threatened. In the bush of southwestern Australia, another arid region of high endemism, about two-thirds of the 1,600 plant species are endangered by the rapid spread of a fungal disease carried inadvertently by humans walking or driving through the bush. And in the United States, about 3,000 plants, nearly one in every eight native species, are considered in danger of extinction; without strong efforts to save them, more than 700 are likely to disappear in the next 10 years.[21]

Some analysts believe, however, that the greatest threat to human welfare comes from losses at the third generic level of diversity—the genetic variety within species, most notably food crops and their wild relatives. Farmers have used and created genetic diversity for millennia to increase agricultural production. Crop breeding and genetic engineering (which can only rearrange existing genes, not create them) are no less dependent on it. In developing nations, especially, generations of farmers have developed a remarkable array of crops. The Ifugao people of the island of Luzon in the Philippines identify more than 200 varieties of sweet potato by name. Farmers in India have planted perhaps 30,000 different strains of rice over the past 50 years.[22]

The widespread introduction of a handful of high-yielding, or "Green Revolution," crop varieties has boosted overall food production over the past several decades but eliminated many traditional strains that were well adapted to local ecosystems and could have been used to develop higher-yielding, locally appropriate crops. If current trends continue, three quarters of India's rice fields may be sown in only 10 varieties by 2005. In Indonesia, 1,500 local varieties of rice have disappeared in the past 15 years, and nearly three fourths of the rice planted today descends from a single maternal plant. Similarly, in the U.S., just six varieties of corn account for 71 percent of the corn fields, and nine varieties of wheat occupy about 50 percent of the wheat land.[23]

Such high levels of agricultural uniformity leave fields vulnerable to pest and disease outbreaks. At their worst, such outbreaks can rampage over entire countries, as in the Irish potato famine of 1846. In 1991, the genetic similarity of orange trees in Brazil provoked the nation's worst-ever outbreak of infections such as citrus cancer, severely reducing output. Fortunately, some farmers add new crops to their traditional menu without completely abandoning the old. Peasants in the Tulumayo Valley of eastern Peru, for example, raise nearly 180 different varieties of potato, even though nearly half of their fields now grow high-yield potatoes.[24]

15

Genetic erosion is a problem among wild, as well as cultivated, life forms. Through population reductions or intentional homogenization, many species have lost much of their internal diversity, and hence their ability to survive collectively. African cheetahs have almost no detectable genetic diversity, and their uniformity has resulted in reproductive problems and vulnerability to disease. On the west coast of North America, according to the American Fisheries Society (AFS), at least 106 major populations of salmon and steelhead have been wiped out and another 214 types of anadromous fish (those that migrate between fresh water and the ocean) are at some risk. Just as monoculture plantations have largely replaced the region's forest wilderness, genetically impoverished hatchery fish have supplanted their wild cousins. About 75 percent of the Columbia River basin's fish are now hatchery-produced and lack the hardiness and survival instincts of wild salmon.[25]

The conclusion of AFS's Endangered Species Committee could easily apply to the biological heritage of much of the world: "With the loss of so many populations prior to our knowledge... the historic richness of the salmon and steelhead resource of the West Coast will never be known. However, it is clear that what has survived is a small proportion of what once existed, and what remains is substantially at risk."

Protecting Ecosystems

For the past century, nature conservation efforts have focused on the protection of habitats in parks and other reserves. This strategy has had an important role in preserving biological diversity. Today there are just under 7,000 nationally protected areas in the world, covering some 651 million hectares, 4.9 percent of the earth's land surface, or 1.3 percent of the earth as a whole. (See Table 3.)[26]

Several nations have, on paper at least, set impressive proportions of

Table 3. Nationally Protected Areas,
Selected Countries, Regions and World, 1990

Country/Region	Area of National Parks and of Equivalent Sites[1]	Share of Total Land Area
	(thousand hectares)	(percent)
Greenland	71,050	32.7
Venezuela	20,265	22.2
Bhutan	924	19.7
Chile	13,650	18.0
Botswana	10,025	17.2
Panama	1,326	17.2
Czechoslovakia	1,964	15.4
Namibia	10,346	12.6
United States	98,342	10.5
Indonesia	17,800	9.3
Australia	45,654	5.9
Canada	49,452	5.0
Mexico	9,420	4.8
Brazil	20,525	2.4
Madagascar	1,078	1.8
Soviet Union	24,073	1.1
WORLD	651,468	4.9

[1]Includes protected areas over 1,000 hectares in IUCN's Categories 1-5 (Strict Nature Reserve, National Park, Natural Monument/Natural Landmark, Wildlife Sanctuary, and Protected Landscape or Seascape); does not include production-oriented areas such as timber reserves or national forests.

Sources: International Union for Conservation of Nature and Natural Resources, *1990 United Nations List of National Parks and Protected Areas* (Gland, Switzerland, and Cambridge, U.K.: 1990); U.N. Food and Agriculture Organization, *Production Yearbook 1989* (Rome: 1990).

their territory off-limits to development. Bhutan, Botswana, Czechoslovakia, Panama, and Venezuela have designated over 15 percent of their lands as parks. National and global figures, however, mask great unevenness. Parks in Chile, for example, are concentrated high in the scenic Andes, and more than half of that country's unique vegetation types are not protected at all. Globally, high-altitude habitats have received a disproportionate share of protective efforts, while others of greater biological significance (such as lowland forests, wetlands, and most aquatic ecosystems) have been neglected. Belize, for instance, protects less than 2 kilometers of its 220-kilometer-long Caribbean barrier reef, the longest in the western hemisphere. The natural systems most commonly found in the world's parks are deserts and tundra.[27]

Although most national parks have been established in the last two decades, societies have been consciously protecting ecosystems for thousands of years. South Asian and Southeast Asian farmers have traditionally honored sacred groves, believed to be homes of powerful deities. The Kuna and Emberá-Chocó Indians of Panama leave patches of old-growth forest as supernatural parks, refuges for both wildlife and spirits. Waterways as well as forests are protected by the Tukano Indians of Brazil, whose taboos guard as much as 62 percent of nearby streams as fish sanctuaries.[28]

These traditional conservation methods, which include both protected areas and diversity-sustaining production systems, appear to have been more effective than their modern counterparts. Although no global tallies of the land area under various indigenous conservation strategies are available, some evidence can be found in a 1989 Sierra Club survey of the world's large wilderness areas—defined as regions larger than 400,000 hectares free of roads or permanent settlements, where major habitat alteration is unlikely to occur. This study found that natural forces, rather than humans, are still dominant over 4.8 billion hectares—more than one third of the earth's land surface. (See Table 4.) By and large, where these wildlands are habitable (the majority are desert and ice), they overlap with the traditional territories of tribal groups. Indigenous people have proved fully capable of abusing land and hunting wildlife to extinction, and it is dangerous to make generalizations about traditional cultures which vary at least as much as their modern counterparts; but it is evident that the world's healthy ecosystems are found predominantly in areas under indigenous control.[29]

Table 4. Estimated Wilderness[1] Areas,
Selected Countries, Regions, and the World, Early 1980s

Country/Region	Area	Share of Total Land Area
	(million hectares)	(percent)
Antarctica	1,321	100
Soviet Union	752	34
Canada	641	65
Australia	229	30
Greenland	217	100
China	211	22
Brazil	202	24
Algeria	140	59
United States	44	5
Peru	37	29
Botswana	31	54
Venezuela	30	33
Angola	27	22
Central African Republic	21	34
Guyana	12	57
Zaire	12	5
Kenya	11	19
Other	869	16
WORLD	4,807	36

[1]Defined as land shaped primarily by the forces of nature; based on surveys for areas larger than 400,000 hectares with no roads or permanent settlements.

Sources: J. Michael McCloskey and Heather Spalding, "A Reconnaissance-Level Inventory of the Amount of Wilderness Remaining in the World," *Ambio*, Vol. 4, 1989; world land area from U.N. Food and Agriculture Organization, *Production Yearbook 1989* (Rome: 1990).

Today, both modern and indigenous conservation systems are unraveling. In many areas, such as the Amazon basin, the often intimate knowledge of nature possessed by indigenous people is fading even faster than nature itself. On average, one Amazon tribe has disappeared each year since 1900. Many more have lost their lands or been assimilated into the mainstream culture. Especially for medicinal plants, traditional crops, and other life forms favored and used by native people, the loss of cultural diversity is one of the greatest threats to biological diversity.[30]

The relationship between traditional cultures and biodiversity is more ambiguous in the case of large animal species. The Dai people of southern China, for example, respect "holy forests," and plant trees to meet their firewood needs, but their hunting is decimating bird populations in the nation's remaining rain forests. Nonetheless, those indigenous practices that do conserve wildlife are vanishing as native cultures and territories succumb to the expanding influence of the global commercial economy. Orangutans, for example, have been virtually wiped out in the Malaysian state of Sarawak, both by hunting and by destruction of their rain forest habitat. Only along the upper reaches of the Batang Ai River in southern Sarawak do they thrive, in part because local Iban people believe it is taboo to kill them. But the Iban, constantly told that their culture is backward, are abandoning their traditional beliefs, and orangutan hunting is reportedly on the increase.[31]

Of the world's nationally protected areas, many, perhaps most, exist largely on paper. In the tropics, most parks have little or no staff or budget, and politically weak or corrupt parks departments are often unable or uninterested in protecting them from damage. A 1988 survey by the Organization of American States found that only 16 out of 100 Caribbean marine parks (outside the United States) had management plans and adequate staff and funding. Similar problems exist in wealthy nations: Shiretoko National Park, Japan's wildest protected area, covers more than 37,000 hectares but has never had more than a single ranger.[32]

In addition, many parks encourage destructive but profitable activities. Logging, for example, occurs with government blessing in parks in Canada, Czechoslovakia, and Indonesia. In Ecuador's Podocarpus National Park, a cloud forest reserve which shelters more bird species than all but two of the world's protected areas, 90 percent of the land has been leased to national and international mining companies in the past

10 years. And in Europe, Canada, and many of the world's marine areas, parks are oriented primarily toward recreation, with biological conservation a secondary or nonexistent concern. Because tourists part with their money more readily than cash-strapped governments, many parks rely on them to pay the bills, often at the price of serious crowding, pollution, and habitat degradation.[33]

The most common threats to protected areas, however, originate in activities outside their boundaries. Severe air pollution blowing into Poland's Ojcow National Park, for example, has helped kill off 43 plant species there, and a third of the remaining plants are threatened. In the Philippines, explosives and cyanide used by local fishers have eliminated more than half the coral cover in Tubbataha Reef National Marine Park in just five years.[34]

In Europe's largest national park, Doñana National Park in the Spanish state of Andalucia, wetlands are being drained of their life-giving underground water by a major irrigation project on the park's borders. Pesticide spills and poaching have also killed thousands of Doñana's birds. Under pressure from the European Community, the Andalucia state government has halted expansion of the irrigation project, and construction of a nearby $100 million tourist complex is also on hold pending study of its environmental impact.[35]

Like its counterparts in the developing world, the Spanish government is demanding compensation from its wealthier neighbors in return for any conservation policies it pursues. While protection of whole ecosystems can offer sizable economic benefits, such as nature tourism revenues and clean and reliable water supplies for downstream communities, it also entails upfront costs. And few benefits reach the people at whose expense preserves are often created. Establishing parks in the tropics has often entailed evicting people or curtailing their customary uses of the area, without compensation. By disrupting traditional management and exacerbating the poverty that often drives environmental degradation, some parks have hastened the devastation of ecosystems they were designed to protect.[36]

Over the past decade, many park managers have come to realize that the

> "Many park managers have come to realize that the survival of protected areas depends ultimately on the support of local people rather than on fences, fines, or armed force."

survival of protected areas depends ultimately on the support of local people rather than on fences, fines, or armed force. A number of promising projects have been initiated by governments and nongovernmental organizations to reduce pressures for exploitation by alleviating poverty in communities in or near protected wildlands. These projects have typically focused on establishing buffer zones around parks where limited exploitation of natural resources by local people is permitted; providing health care, water supplies, or other community services to compensate for lack of access to park resources; and supporting local economic development efforts such as tourism, wildlife ranching, or marketing of non-timber forest products.[37]

The underlying principle of these efforts is sound: poverty alleviation is the only feasible means of protecting natural areas in the long term. But the difficulty of this approach should not be underestimated. Many villagers are rightly suspicious of parks departments which, in their view, have stolen resources from them not for protective purposes but for wealthier bureaucrats to exploit. Park managers typically find calls for more participatory management a threat to their authority. And by no means are all local people, indigenous or otherwise, interested in conservation of natural areas. For example, while the Gwich'in people of Alaska and the Yukon firmly oppose oil drilling in Alaska's Arctic National Wildlife Refuge because of the risk it poses to their caribou-based economy, the neighboring Inupiat Eskimos are among the most vocal supporters of opening the refuge to oil exploration.[38]

Given the opportunity, however, local communities can be powerful forces in the defense of protected areas. The Inter-Ethnic Association for the Development of the Peruvian Amazon, a coalition of 27 indigenous groups, has actively opposed the government's push to open the 2-million-hectare Pacaya-Samiria reserve to oil exploration. Along with Peruvian environmentalists, they insist that national law—which forbids such activities in protected zones—be upheld, and that any drilling take place outside the 6 percent of the Peruvian Amazon ostensibly protected as parkland. The indigenous coalition is especially concerned about potential contamination of the waters and fisheries used by the 15,000 people who live along the rivers in the reserve, the nation's largest protected area.[39]

A recent survey of 23 projects aimed at improving park protection and local economic well-being in Africa, Asia, and Latin America concluded that they have had limited impact to date. According to the authors, World Wildlife Fund Senior Fellow Katrina Brandon and World Bank consultant Michael Wells, most projects are only a few years old, too small to cope adequately with the problems they seek to address, and unable to establish explicit linkages between conservation and development. For example, local people now receive the gate fees paid by butterfly-watching tourists in Mexico's Monarch Overwintering Reserve. But this alternative source of income has not deterred them from logging in the area.[40]

Given the great difficulties in protecting whole ecosystems, people representing a wide range of interests have called for solutions less demanding than the defense of wilderness—defined as very large, roadless, lightly managed, minimally polluted ecosystems. Scientists have come to realize that change and disturbance—from fire and windstorms to the ancient practices of indigenous peoples—play important roles in fostering biological diversity. And foresters, among others, are beginning to design production systems aimed at mimicking these processes. In addition, it can be argued that there are no truly natural areas on the planet, since the impacts of indigenous people and industrial pollutants have been found in even the remotest territories.[41]

Why, then, maintain large areas with minimal human use? Quite simply, no other approach can protect biodiversity—at all its levels of organization—as well. Grizzly bear, harpy eagle, hornbill, spotted owl: species that roam over large areas, that require specialized habitats, or that do not get along well with humans need large wilderness reserves if they are to survive outside of zoos. In the Amazon, woolly monkeys (which play a key ecological role by dispersing seeds too large for other animals to eat) shy away from roads and clearings; they are severely threatened by hunters in any area occupied by either Indians or non-tribal settlers. As American ecologist Reed Noss writes, "No matter how ecologically sophisticated our land management practices, how perfectly we think we can mimic natural processes, so long as there are roads and mechanized humans, some species will disappear."[42]

Because disturbance is essential to natural ecosystems but often inconvenient to humans, large areas of wildlands are needed to let fires and other large-scale natural phenomena play their course uninterrupted. Also, for protected areas to be functional over the long term, they have to be ample enough to ensure that only a fraction of the area is involved in any given disturbance. Habitats too small to weather the impacts of random variation in weather and wildlife populations will not be able to maintain their full complement of genes, species, and functions. Where smaller parcels of habitat are all that remain, more active (and costly) management will be required.[43]

If exploitation is minimal, areas of sustainable resource extraction can be nearly indistinguishable from wildlands. In these places, maintaining ecological integrity means continuing the traditional uses of the ecosystem. But this should not be confused with the more intensive commercial uses sometimes proposed as "sustainable" alternatives. The difference may be subtle, yet critical. For example, the widely hunted uacari monkeys of the Amazon basin have survived near Tukano Indian villages, which are one or two days' paddling apart, but among more closely spaced villages they are hunted close to extinction. Intensifying the use of minimally exploited areas will undoubtedly bring about biological losses.[44]

In wilderness-rich nations such as Brazil, the Guianas, or Papua New Guinea, it is inevitable that many natural habitats will be converted as economies and populations grow. This is no excuse, however, for continued government-sponsored depredation of wildlands when there are large, often more suitable areas of degraded land available for the same purpose. In Latin America, for example, 1 percent of landowners commonly control more than 40 percent of arable land, much of which stands idle on huge estates. Here as in much of the tropics, the best hope for deflecting pressure off intact ecosystems and for bettering the lot of the rural poor lies in reforming the skewed patterns of resource ownership that lock up the best resources in the hands of a tiny minority.[45]

In most nations, the opportunity to achieve a balance between wild and domesticated landscapes was lost long ago, and a necessary (if seemingly Draconian) goal is to cease "developing" any more relatively undamaged ecosystems. A first step toward this is determining where these areas are. In the United States, an Endangered Ecosystems Act has been proposed that would inventory and restrict development of any ecosystems reduced below a given percentage. On a global scale, Israeli landscape ecologist Ze'ev Naveh has proposed the compilation of a list of threatened landscapes.[46]

Because so many remaining areas of global importance to biodiversity are inhabited by indigenous people, who usually have the greatest knowledge of the ecosystem to be safeguarded, protection will need to be on their terms. A first step is the recognition of the rights of people to lands and resources they have used for generations. Despite a flurry of recent homeland recognitions in the Amazon basin, Native Americans have legal rights to less than 5 percent of their ancestral domains in the Americas, and Southeast Asia's forest dwellers have been similarly dispossessed. Where current or desired practices are incompatible with biodiversity conservation, governments or conservation advocates can negotiate with traditional residents about the protection of ecosystems and compensation for any benefits they forgo.[47]

In South Africa, infamous for its racially skewed distribution of land and power, the parks department recently agreed with local herders to establish Richtersveld National Park, the nation's first "contractual" park. The Nama pastoralists of the area have agreed to herd their sheep and goats under new restrictions; in return, they get a healthy portion of all earnings from the park and are well represented on its management committee. Similar efforts are under way to establish cooperative protected areas in Papua New Guinea, American Samoa, and Western Samoa.[48]

Beyond allowing greater local input into the designation and management of protected areas, the root causes of degradation must be addressed. Usually agents rather than causes of destruction, the rural poor face complex forces pushing them to harmful short-term behavior; turning back these forces, and the equally serious vectors of degradation in industrial nations, will take far-reaching changes in government policies, corporate behavior, distribution of resources, and individual ethics. Fundamentally, it will require an embrace of diversity not just in the areas set aside from human dominance, but in those we dominate as well.

> "Native Americans have legal rights to less than 5 percent of their ancestral domains, and Southeast Asia's forest dwellers have been similarly dispossessed."

Conservation Beyond Parks

Even if societies undertake radical efforts to curb their impacts on remaining natural areas, most of the world's landscapes will continue to be dominated by human activity, and most will fall outside the scope of strictly protected reserves. Semi-natural areas—such as second-growth forests, waters whose fish are intensively harvested, and rangelands grazed by livestock—prevail around the globe. Unless society can learn to tolerate and maintain wildness in these civilized landscapes, biodiversity has a bleak future.

Untrammeled nature has long been viewed as an obstacle to economic development, and ecological decline as the inevitable price of progress. The protection of ecosystems within parks has usually been predicated on the assumption that natural systems outside them could then be homogenized at will. Recently, however, both environmentalists and industrialists have espoused the idea that economic and environmental well-being are linked, not opposed, and have begun to seek ways to blend production and protection.

Reasoning that diversity must be economically rewarding to rural people if it is to survive, nonprofit groups and green-minded businesses are bringing new foods, cosmetics, medicines, soaps, and other products from the world's tropical forests to stores in Europe and North America. The range of products hidden in forests, reefs, and other ecosystems is indeed a powerful argument for their conservation. Nature is the world's greatest chemistry laboratory, and the evolutionary struggle has endowed many creatures with chemical defenses that are a largely untapped source of potentially useful compounds. For example, less than 1 percent of the plants of Madagascar, where medicinal plants are a major export as well as the basis of local health care, have been analyzed chemically. And animals as well as plants might help save human lives. Drugs derived from anticoagulants produced by snakes, ticks, and vampire bats may soon help prevent clogged arteries. Other natural substances may prove useful as antidotes to poisons; Madagascar's golden bamboo lemur, which is endangered, is remarkably able to detoxify the cyanide found in large amounts in the bamboo shoots it eats.[49]

In 1991, environmentalists' argument that biological conservation makes economic sense received a strong boost when Merck & Co., the world's largest pharmaceutical firm, signed an agreement to pay Costa Rica's National Biodiversity Institute (INBio) $1 million for the right to screen insects, microbes, and plants for their medicinal properties. INBio's crew of parataxonomists will gather and catalog species from Costa Rican forests, and royalties on any drugs that are developed by Merck out of the collaboration will go to INBio, with the funds earmarked for conservation projects. While it remains unclear how much local people will get to benefit from and participate in the project (beyond being hired as specimen gatherers), the venture makes tangible the potential returns of saving and studying living things.[50]

In discussions of the *potential* benefits of biodiversity to humanity, however, it is frequently overlooked that many species little known to science or industry are already used and managed by local communities. Tropical forests are critical to hundreds of millions of rural people as sources of nutrition, health care, raw materials, and cash income. The diverse species of the tropics form a major part of Third World economies and are especially important for the rural poor, who often pay the direct costs of biological degradation. Four fifths of humanity relies on traditional medicine (based largely on tropical plants) for their health care, while rain forest plants are the source of key ingredients (now usually cultivated or synthesized) in pharmaceuticals worth tens of billions of dollars annually. In northwest Amazonia alone, some 2,000 species are used for medicinal purposes. International commerce in the most widely traded nontimber forest product, rattan (palm stems used for wicker furniture and baskets), is alone worth roughly $3 billion annually.[51]

Forests, rangelands, fisheries, and wildlife throughout the developing world have long been managed as common property resources, often with few negative ecological impacts. For example, in eastern Africa, an area famous for its game reserves, large wildlife species are usually more common outside parks, where pastoralists' activities over the ages have helped maintain the savannahs, than inside the parks, where traditional herders have been removed. Many communal management systems, however, are unraveling as populations surge, traditional cultures erode, and national governments confiscate or privatize resources held by communities. In addition, when subsistence-level economies have adopted modern technologies or become commercialized, increasing levels of production have tended to strain local ecosystems without improving local welfare.[52]

From Southeast Asia's rattan to the fish and fruits of the Peruvian Amazon, the usual fate of species that gain long-term popularity in industrial markets is depletion. And the usual lot of their harvesters is continued poverty, as the profits from their work are siphoned off by powerful intermediaries and distant elites. Brazil nut gatherers, for example, receive about 4 cents a pound for their labors, just 2-3 percent of the New York wholesale price. Three fourths of the Brazilian market is controlled by three companies, owned by three cousins.[53]

Brazil's rubber tappers are perhaps the most renowned of the world's forest people in their efforts to break out of this cycle of human and environmental suffering. Though most are chronically in debt and under threat of violence or death from landowners, they have gained some government support for their claims against cattle ranchers and speculators who seek to clear the forest. Their vision of an alternative form of development has centered around extractive reserves: areas owned by the state but managed by local communities for uses compatible with the continued existence of the forest. Since 1987, the Brazilian government has recognized 14 extractive reserves covering 3 million hectares. The national rubber tappers' union hopes eventually to manage 100 million hectares (one fourth the Brazilian Amazon) in this way. These successes have inspired similar efforts in other countries, including the declaration of forest and lake reserves in Guatemala and Peru.[54]

Nonetheless, Brazil's forest extractors still face serious economic, ecological, and political hurdles. Although they hope to diversify their economic base, they currently get most of their income from rubber and Brazil nuts, the economic viability of which is uncertain at best. Extractive reserves have difficulty competing with their monocultural competitors. Until 1990, rubber prices in Brazil were propped up to three times the world price with heavy taxes on imports of cheap rubber from Asian plantations. When the props were removed, the prices fell by two-thirds. Single-species stands of Brazil-nut trees have been started in Malaysia, and a decade-old stand near Manaus, Brazil, is expected to begin producing a commercial crop soon. As with other Amazonian plantations, disease will likely besiege the dense rows of domesticated trees, but perhaps not before they have flooded the market with low-priced nuts.[55]

Tapping trees and gathering fruit are fairly benign activities, especially compared with the wholesale destruction usually caused by cattle ranching or commercial logging in tropical forests. But removing industrial quantities of any material from an ecosystem is likely to cause major changes. Although traditional people's knowledge of their local area usually outstrips that of biologists, neither group has much experience in sustainable commercial use of tropical ecosystems. Biologists know little or nothing about most of the species of the tropics; even the ecology of commercially important species remains little understood. No one knows, for example, if the pronounced absence of Brazil nut seedlings and saplings in many areas today is a natural cycle or a result of over-harvesting.[56]

Rain forest trees form the base of a complex food chain. In turn, they rely on birds, insects, mammals, and even fish to disperse their pollen and seeds. According to ecologist Charles M. Peters of the New York Botanical Garden (NYBG), "it is obvious that removing commercial quantities of fruits, nuts, and oil seeds will have an impact on local animal populations." The economic incentive to "enrich" forests with money-making species and cut down their competitors could also reduce the diversity of plants (along with symbiotic animals) within extraction areas. Without careful management, losses at the genetic level are also possible: as the best fruits or leaves from a given population are continually harvested, the less desirable plants may be the ones to survive over time.[57]

The dangers of market-oriented extractive economies can be minimized by carefully choosing species and ecosystems. Paradoxically, a habitat can have too much biological wealth to serve as a sustainable source of income on a commercial scale. Tropical waters, for example, are home to a rich variety of fish; but the rarity of most individual species makes the harvest levels demanded by modern high-tech fleets difficult to sustain. Similarly, tropical moist forests are home to a great array of vertebrates, but the relatively low densities of most species make them vulnerable to decimation by any major commercial pursuit. Savannahs and grasslands, in contrast, contain high densities of large-bodied, rapidly reproducing animals and are thus more promising sites for the commercial use of wildlife. Diverse habitats lend themselves better to those who fish, hunt, or gather a diversity of species.[58]

Within the tropical forest realm, some areas are naturally less diverse than others and thick with commercially promising species, such as the vitamin C-rich fruit of the *camu camu* that grows in dense stands throughout the floodplains of the Peruvian Amazon. Ancient yet semi-natural forests also hold great promise. Anthropologist Christine Padoch of the NYBG reports finding fruit-filled, remarkably diverse groves (44 tree species within one-fifth of a hectare) in Kalimantan, Indonesia, that had been created by generations of villagers casually planting, weeding, and even spitting out fruit seeds over their shoulders.[59]

In some cases, making the use of diversity rewarding to local people can be aided by developing markets for new or unknown products. But this approach will only work if ecological limits are observed and if the fundamental concerns of the rural poor are addressed. Political reforms will be required to secure rural people's rights to land and resources, reduce their vulnerability to exploitation and violence by outsiders, and ensure that the benefits of conservation stay within local communities.

In Africa, wildlife is typically considered state property (a legacy of the colonial era), and benefits from hunting or viewing animals usually flow to national coffers and foreign tour operators. Local peasants have borne the costs of living with wildlife, such as trampled crops and attacks on livestock, with no recompense short of poaching for the black market.[60]

In several southern African nations, the state has ceded exclusive control over wildlife. Rural communities living in 10 of Zambia's 31 Game Management Areas have been granted rights to wildlife; poaching has fallen dramatically and wildlife populations appear to be rebounding as a result. Area youths, who know the wildlife and local bush better than anyone else, are now trained as village scouts to keep out poachers; revenues from safari hunting fees go into a revolving fund that supports both conservation activities and community development projects. In rural Zimbabwe, another program of joint state-community management has quickly gained popularity: as of 1990, residents of communal lands covering nearly one eighth of the country had begun, or were planning to begin, participatory wildlife management programs.[61]

Besides fulfilling the legitimate historical claims of local people, restoring some degree of local control over resources is probably the only way that vast areas in the tropics can be "managed" at all. Governments claim ownership of 80 percent of the world's remaining mature tropical forests, but have little hope of controlling how they are used. Each officer in Indonesia's Ministry of Forestry, which has a full-time field staff of 17,000, is responsible for more than 8,000 hectares of state forest. Meanwhile, more than 30 million Indonesians live on or near the 143 million hectares of state forest land, an area the size of Alaska. The nation also has the world's longest coastline and more than a million fishers, yet the government claims all responsibility for fisheries management. Cost as well as staffing constraints favor new approaches. Zambia's joint management of wildlife with villagers is more effective and costs less than 10 percent as much as the law enforcement approach favored in much of Africa.[62]

Full recognition of traditional resource rights, however, is not sufficient to ensure sustainable use of those resources. In Papua New Guinea, native people have full ownership of 97 percent of the country's largely forested land. But predatory logging by Japanese and other firms is no less a problem here than in Malaysia, where the state (along with closely allied timber companies) controls forests. The Papuan state is too weak and corruption-ridden to prevent widespread abuses by multinationals, and local owners are easily swayed into selling their forests at discount rates.[63]

Reconciling diversity with development will require loosening the grip of industry on natural resources and accepting that production of any single commodity will be constrained if ecosystems are managed for

> "Governments claim ownership of 80 percent of the world's remaining mature tropical forests, but have little hope of controlling how they are used."

diversity. The success of some U.S. foresters' efforts to develop a "new forestry," for example, depends on whether they can free forest management from its historical timber bias. Rather than homogenize ecosystems in order to maximize wood production, new forestry ostensibly aims to maintain and use the ecological complexity of natural forests. But because diversity conservation will require protecting the few scattered remnants of pristine forest, and logging other areas less intensively, new forestry cannot succeed without major reductions in the timber harvest taken from national forests—a politically charged issue.[64]

While new forestry, extractive reserves, and other programs of sustainable commercial use of ecosystems are in their infancy, it is already clear that their success depends on programs of sustainable *nonuse*. Protecting large areas of undisturbed forest near logging or extraction areas is the only way to ensure the survival of animals that pollinate and disperse the seeds of commercially important tropical trees. Studies in the western United States have shown that intact stands of native forest harbor insect predators which safeguard nearby plantations from pest outbreaks. Similarly, the protection of coral reefs from fishing in Belize and the Philippines has revived local fishing economies by providing depleted fish stocks room to grow. Ultimately, only natural areas can tell us if our uses are in fact sustainable. As essayist Wendell Berry writes, "we cannot know what we are doing until we know what nature would be doing if we were doing nothing."[65]

Living with Diversity

Fishers' nets and loggers' saws may directly impoverish local ecosystems, but most biological losses have root causes far away, in long-settled urban areas and farms where diversity is seldom a concern, but where steadily rising demand for food, water, wood and other resources—and the dispersal of resulting wastes—reach far beyond the settled areas themselves. In general, these peopled landscapes have lost much of their own biological wealth, but what remains is still important to their continued functioning and livability. Reconciling farms and cities with diversity will require stopping the damage they bring to remaining natural habitats, but also beginning to halt and reverse the homogenization of these *unnatural* habitats.

Uniformity is not inherently undesirable. In fact, to some degree, homogeneity is the basis of all agriculture: a given type of plant is favored and others are suppressed or eliminated. But trends in recent decades (most notably the Green Revolution and the parallel intensification of farming systems in industrial nations) have pushed uniformity to dangerous levels.

The unsustainability of modern agriculture is in part a measure of its inability to tolerate diversity. Both genetic and ecological uniformity—the sameness of fields sown horizon to horizon without interruption—demand costly and often futile reliance on chemicals to protect crops from pests or diseases that are rapidly spreading and evolving. The drive to leave no hectare unplowed worsens soil erosion, pushing tractors onto highly erodible hillsides and removing windbreaks, hedgerows and other remnant habitats.[66]

Some of agriculture's biological impacts are obvious—the expansion of farms onto forests and wetlands, for example. While the increasing reliance on chemical inputs and machinery has reduced these impacts in some cases by decreasing the area needed to produce a given amount of food, it has worsened others. Chemical runoff and soil erosion from farms on the Belize coast, for example, constitute one of the greatest threats to that nation's barrier reef. Freshwater as well as coastal habitats in the United States and Europe suffer acutely from the runoff of fertilizers and pesticides from croplands. In the United Kingdom, many bird species have declined over the past 20-30 years, likely due to increased herbicide use and the elimination of hedgerows and unused lands from agricultural landscapes.[67]

A fundamental transition away from today's wasteful and polluting farming systems is needed to put the world's food supplies on a secure footing. Many of the reforms that will reduce farming's dependence on fossil fuel inputs and its misuse of soils and waters can also restore diversity to agricultural landscapes. Pesticides, for example, kill not only pests but other animals, such as pollinators and predators, that are bene-

"Protection of habitat is the single most effective means of conserving biological diversity."

ficial to agriculture. Alternative pest control measures that lower pesticide use can also, ironically, reduce pest damage to crops by reviving the diversity of soil and insect communities, which play crucial roles in maintaining soil productivity and checking the spread of pest outbreaks.[68]

The likely reduction of agricultural subsidies in Europe under international trade agreements could cause large areas of farmland to fall out of production, including many traditional farms that have considerable value as wildlife habitat on a continent largely devoid of pristine nature. But proper planning could ensure that highly erodible cropland and ecologically sensitive areas such as wetlands are retired first, in order to reclaim vital habitats and protect waterways. The U.S. government spends billions of dollars each year to idle croplands for price-support or erosion-control reasons; these programs could easily encourage the planting of wildlife-supporting vegetation or the maintenance of biological corridors—crucial links between the country's tattered remnants of natural habitat.[69]

Many *ex situ* (off-site) efforts, including gene banks, botanical gardens, and zoos, are necessary, but protection of habitat is the single most effective means of conserving diversity. This is no less true for traditional crop varieties, whose habitats are the world's small farms. Protecting these dynamic habitats means reversing the policies and forces that push farmers off the land or encourage them to abandon customary crops. Some of these forces (such as subsidies for pesticides, and top-down agricultural research programs that scorn local knowledge) are largely hidden. Less hidden is the highly skewed ownership of fertile land in most developing countries, that pushes peasants into urban slums or fragile ecosystems unsuited for cultivation.[70]

A cause of genetic erosion that has received more attention is the lack of legal recognition and compensation for traditional people's contributions to genetic diversity. Historically, plant breeders, biotechnology firms, and pharmaceutical companies have had unrestricted access to the developing world's genetic diversity, which has been considered a "common heritage of mankind" regardless of the innovative role of individuals in managing or creating a given variety. Conversely, the indus-

33

trial users of biodiversity have been able to protect and profit from their technological innovations through patents and other legal protections. Beyond the obvious injustice of what can only be described as a form of theft, local people denied the benefits of their work in conserving and breeding genetic varieties have little economic incentive to continue it.

How small farmers, medicinal healers, and others should be compensated for the use of their knowledge (which is often held collectively rather than by individuals) is a controversial matter. Discussions are under way in several international forums regarding mechanisms of recognition and compensation for intellectual property rights. Nations currently negotiating a global biodiversity convention appear to have come to a consensus rejecting the "common heritage of mankind" approach in favor of increased national sovereignty over genetic resources, a move that conservationists generally welcome. But the more complex issue of ensuring local, rather than national, rights to genetic resources has not been broached in most international discussions. Moreover, the reluctance of Northern nations to provide the funding or technology necessary for Southern nations to maintain their biological wealth threatens to stall biodiversity negotiations altogether.[71]

Traditional agroecosystems are important not only because they provide sustenance to rural people and harbor valuable genetic resources, but also because they contain the seeds of a sustainable, diversity-based mode of agriculture. At varying levels, diversity is the basis of production for many peasants. Farmers often mix strains of a given crop in their fields as a hedge against the vagaries of weather. They also tend to recognize the dependence of their farms on adjacent ecological systems and to tolerate wild plants (often crop relatives whose continued interbreeding with domestic descendants contributes to genetic variety) on the outskirts of their fields. More commonly, farmers mimic the species diversity of natural systems and grow varied crops together; such multiple cropping still accounts for as much as 20 percent of world food production.[72]

Population growth and the expansion of large commercial farms have rendered many once-sound practices no longer viable, and traditional agriculture badly needs infusions of money and research to increase its modest yields without abandoning its stability. Few generalizations can be made about the variety of sustainable approaches to agriculture. But to one degree or another, diversity and information replace energy and chemicals as the primary basis of such systems. As much as any gene bank or biotechnology firm, it is the traditional farms of the world that contain the healthy soil biota, patches of natural vegetation for pollinators, polycultures, and other types of diversity—and the locally relevant knowledge—that make sustainable agriculture possible.[73]

Urban areas, with good reason, are considered the antithesis of natural diversity. Only the most resilient creatures (many of them regarded as weeds and pests) thrive in them, and cities' ceaseless expansion, consumption of resources, and emissions of waste threaten both farmland and wilderness almost everywhere. As with agricultural lands, the first priority for urban areas is to halt their expansion onto other ecosystems and reduce the damage they export, such as the sewage poured onto coral reefs by burgeoning coastal cities throughout the tropics, or the wasteful consumption of tropical hardwoods in Japanese building construction.

But even concrete jungles can support some diversity. Landscaping of private yards and public spaces with native vegetation can not only reduce the expense and environmental impact of watering, spraying and hauling the remains of sterile grass monocultures, but also help revive bird and other wildlife populations. Most urban areas also have waterways running through them, or corridors of unused land such as steep ravines; if their use as waste receptacles is reduced, these can be maintained or restored as wildlife habitat. A number of European and U.S. cities have moved to establish greenway corridors along rivers, old railroad tracks, and urban perimeters. Although created and managed primarily for recreational purposes, these greenways have great biological potential to help reconnect increasingly fragmented and dysfunctional enclaves of nature beyond city limits.[74]

In developing nations, especially, a surprising amount of agricultural production takes place within city limits, in home gardens. These hidden farmlands contain a great deal of genetic diversity, and their expansion could help reduce the scale and environmental impacts of commercial agriculture. In the United States, converting a quarter of the country's 13 million hectares of lawn to gardens would be equivalent to expanding its cropland by 2 percent.[75]

One reason that the destruction of biological diversity has gone so far without major public commitments to stopping it is that urban dwellers have little experience of the natural and even less understanding of its importance. Restoring nature where people live—reestablishing a personal link with the living world—may be necessary to save it elsewhere. For all the rational arguments favoring long-term protection of biological assets, people who have lost all direct sense of their dependence on natural systems may simply not care.

Only a growing respect for diversity for its own sake—beginning, perhaps, with a reconnection between people and nature within the urban environment—will trigger altruistic responses among those wealthy enough to have the option of considering the needs of future generations and natural communities. Although many conservation measures make economic sense, arguments of economics or self-interest will likely fail to be convincing when the contest is between a few uncharismatic species of unknown value and a major industrial project. As ethnographer Eugene Anderson notes, "Human beings make sacrifices for what they love." Those who maintain strong bonds with the biological world on which they depend may be more inclined to make the hard decisions needed to protect it.[76]

Greenhouse Biology

Some 2.4 million years ago, the earth's climate was cooling and the mammals of eastern Africa faced a challenge. Forests were shrinking and habitats being altered by the changing atmosphere. According to paleontologists' theories, the mammals responded over many millennia by evolving into entirely new species, and whole new families of life were born, including the genus *Homo*, the human beings. Today, the only survivor of that lineage—*Homo sapiens*, our own kind—has so altered the atmosphere that a global warming, far too rapid for most plant or animal species to adapt, now seems likely.[77]

In the past 150 years, humanity has burned enough fossil fuels and vegetation to increase the amount of heat-trapping carbon dioxide in the atmosphere by 25 percent. The U.N.-sponsored Intergovernmental Panel on Climate Change (IPCC) predicts that the continuing emissions of carbon dioxide and other greenhouse gases are likely to raise the planet's average temperature roughly 0.3 degrees Celsius per decade, or 1.5 to 4.5 degrees over the next century. Such increases may sound small, but they would take civilization into uncharted territory. A 3-degree warming would raise the earth's temperature to its highest level in 100,000 years; a 4-degree increase would make the earth warmer than it has been for 40 million years.[78]

The magnitude and rate of global warming in store remain uncertain; the impacts of warming on biological diversity are even less predictable. But a safe conclusion from available information is that without major and immediate reductions in greenhouse gas emissions, the impacts of global warming will probably make the world's current biological collapse pale in comparison.

A warming world will be hostile to life in countless ways. Most notably, natural communities will be forced to migrate away from the equator, or up from sea level, if they are to maintain their usual climatic conditions. For most species and habitats, which have dealt with changing conditions for eons, the warming itself is less a concern than the rate at which it is likely to occur. A century's worth of warming at the IPCC-predicted rate (0.3 degrees per decade, with much greater warming in Arctic and Antarctic regions) would displace vegetation zones in the world's middle latitudes roughly 300 kilometers away from the equator, or in mountainous regions 500 meters higher in altitude. By comparison, many North American tree species moved 10 to 40 kilometers per century as the last ice age ended 10,000 years ago. The predicted warming, therefore, is likely to result in large-scale diebacks of beech, oak and other deciduous forests. To the north, according to some studies, 40 percent of the world's boreal forests (the woodlands of the far North that cover an area nearly the size of South America) may be killed off by climate changes induced by a doubling of CO_2 concentrations.[79]

Warming will be less near the equator than at higher latitudes, but tropical forests will not be immune to the effects of climate change. Shifting rainfall patterns are expected to disrupt the flowering and reproduction of many tropical plants, thereby affecting many fruit-eating animals. Changes in the timing of rainfall could also make forests more vulnerable to their usual mode of destruction—fires set by farmers and cattle ranchers.[80]

While ecologists talk of forests or other vegetation types moving smoothly across the landscape, in actuality, species respond individually to changing conditions. Thus, species currently occupying the same habitat may adapt (or fail to adapt) in different ways. Rather than entire communities migrating together, communities will be disrupted and new ones created as their components disperse at different rates and in different directions.

In this process, there may be winners as well as losers. Creatures that reproduce rapidly and thrive in recently-disturbed habitats (often called weeds or pests) may expand their ranges at the expense of those that reproduce slowly or require mature or old-growth communities to thrive. One study predicts that sea grass beds (important fish and waterfowl habitats found along the coastlines of every continent) will expand under CO_2-induced global warming. But oceanic ecosystems as a whole are likely to be greatly disrupted by the combined effects of climate change, even though they will warm more slowly than their on-land counterparts. (See Table 5.) And since disturbances such as storms, heat waves, and fires may become more frequent and severe with global warming, predictions of biological benefits based on changing *average* conditions are probably flawed.[81]

The majority of communities and species are likely to suffer from the pace and magnitude of change, and some are especially at risk. Species that inhabit oceanic islands and freshwater habitats on land will generally be unable to migrate across the inhospitable environments that surround their local habitats. They will have to survive climate change in place. Similarly, mountaintop habitats and those at the poleward edge of continents have nowhere to move. Tundra ecosystems—home to relatively few species but important for many migratory birds—could be reduced throughout North America and Eurasia, releasing CO_2 and the potent greenhouse gas methane as their soils warm.[82]

Table 5. Likely Impacts of Global Warming on Oceanic Biological Diversity

Predicted Change	Likely Effect
Rising Atmospheric Temperatures (1.5-4.5°C by 2100)[1]	Water temperatures only 2 degrees above normal can cause corals to bleach. Widespread bleaching and die-offs of coral reefs are expected.
	Arctic ecosystems will be disrupted by the increased melting of sea ice in the spring. Overall biological productivity is likely to decline sharply, with marine mammals and seabirds especially affected.
	For species not adversely affected by other changes, rising temperatures may increase growth and metabolism rates.
	Long-distance migrations of animals such as shore-birds and whales may be disrupted as regions farther from the equator warm more rapidly than tropical latitudes.
Rising Sea Levels (30-110 centimeters by 2100)[1]	Most mangrove ecosystems are expected to collapse.
	In the absence of stresses caused by warmer waters, increased wave action or ozone layer depletion, fast-growing coral types may thrive under moderately rising seas.
Shifting Currents (unpredictable)	Most marine animals spend part of their lives as free-floating larvae, dependent for future growth on being carried by currents to suitable habitats. Shifting currents, caused by increasing temperatures, could prevent many from reaching their desired habitats.
	Weakening of upwelling currents (which bring cold, nutrient-rich waters toward the surface) could displace fisheries of major economic importance, such as those off Namibia and Peru.
Increasing Storms (unpredictable)	Increased wave action and storms could directly erode corals, wetlands and other coastal ecosystems and lead to greater flooding of coastal areas.

[1] Intergovernmental Panel on Climate Change predictions, 1990.

Source: Worldwatch Institute, based on sources documented in endnote 81.

As atmospheric and oceanic temperatures rise, the seas will expand and, it is believed, lend more energy from their surface waters to tropical storms and waves. According to IPCC predictions, mean sea level will rise between 30 and 110 centimeters by the year 2100. Combined with more forceful and frequent hurricanes, rising sea level could inundate tens of millions of hectares along the world's coastlines.[83]

Some coastal habitats should be able to migrate along with the rising tide, but the pace of change will overwhelm others' ability to adapt. Mangrove ecosystems (or mangals), the widespread wetlands of the tropics, can keep up with seas that rise less than 9 centimeters per century, but most become stressed during rises of 9 to 12 centimeters per century, according to a recent study by geographers Joanna Ellison and David Stoddart of the University of California. They conclude, "the predicted possible rates of greenhouse-induced sea-level rise of 100-200 centimeters [per] 100 years make it inevitable that most mangals will collapse as viable coastal ecosystems."[84]

Of course, rising temperatures and tides will happen at a time when many of the world's biological systems are already degraded and more susceptible to additional stresses. Wildlife species or agricultural systems that have little genetic diversity will be ill-suited to respond to change. And ecosystems that have already been badly degraded or robbed of species are more likely to succumb to threats that a healthy system might handle. For example, the seeds of oaks, hickories, and other nut-bearing trees of North America may have no means of moving rapidly northward because the passenger pigeon, whose great flocks once darkened skies and dispersed seeds across the continent, was hunted to extinction at the beginning of this century.[85]

Even if reduced emissions of greenhouse gases slow global warming to a pace that migrating ecosystems can keep up with, habitats blocked by barriers such as mountain ranges or steep coastlines will be unable to move. And those fragmented by roads, clearings, and human settlements will literally be backed to the wall. According to the U.S. Environmental Protection Agency, even with the removal of beachfront developments to allow the inland migration of the coastal zone, 30 to 70 percent of the nation's coastal wetlands would be lost to a 1.5 to 2-meter rise in sea level, because of the topography of most U.S. shorelines. If efforts are made to stem the tide with levees and bulkheads, 50 to 80 percent would be lost.[86]

Protected islands of nature in seas of degradation will also fail to withstand the tide of moving species. While few of the world's parks are big enough to support their populations of large mammals, under global warming, many will not even be able to support their plants. Patrick Halpin, an ecologist at the University of Virginia, studied the vegetation zones of 243 U. N. Biosphere Reserves around the world and found that, depending on which climatological models for global warming are used, local climates in 55 to 80 percent of these world-class reserves would no longer be suitable for the current vegetation during the next century. And a recent study of 2,600 large nature reserves (over 1,000 hectares) worldwide predicts that the vegetation in up to 60 percent of these reserves (again, varying by the climatological model used) may be displaced by global warming.[87]

Long-term biological conservation will be impossible without rapid reductions of greenhouse gas emissions, achieved through such measures as energy efficiency and conservation, a shift to renewable energy sources, and the protection and restoration of forests. Unfortunately, these efforts alone will not keep the earth tolerable for its natural and human communities, either. Because of the greenhouse gases already in the atmosphere, a one-degree warming—sufficient to cause major biological disruption—appears inevitable during the 21st century.[88]

Biological diversity can only be safeguarded by planning now and maintaining options for an uncertain future. Site-specific research remains crucial to predicting the impacts of global warming and to learning how to help ecosystems and species adapt. But some measures, such as the protection and restoration of wildland corridors, particularly those running north-south, uphill, and inland from coastal wetlands, are clearly essential for uninterrupted migrations of plants and animals. Flexible management of buffer zones around protected areas can also help accommodate the shifting of habitats. Land-use planning, which usually focuses on the management of particular parcels of land or water but not on the overall landscape, needs to incorporate this broader regional perspective. Such a widening of perspective is especially important in the ocean realm, where currents and wide-ranging species connect disparate regions into coherent ecological units.[89]

Instead of pushing oceanic fisheries, timberlands or other semi-natural areas to the limit by seeking to extract what is termed the "maximum sustainable yield," managers would be well-advised to leave a margin of flexibility to adjust to climatic or other shocks. (And since the ability of natural systems to tolerate long-term exploitation is usually only superficially understood, leaving room for error is desirable on its own merit.) Reducing existing stresses by managing for resilience rather than maximum yield will enable ecosystems and associated human institutions to better cope with pollution and the stress of climate change.[90]

The specter of global warming makes even clearer the importance of incorporating diversity into all human production systems, as well as protecting it in wilderness. Resilient, diverse agricultural systems and semi-natural ecosystems with as full a complement of their native species and interrelationships as possible are more likely to cope with the stresses of a changing climate. The array of crops within a traditional farmer's field is likely to yield a decent harvest in most kinds of weather, and since traditional varieties have been bred to thrive in difficult agricultural environments, they will become increasingly valuable as sub-ideal conditions spread in a greenhouse world. Whatever use is chosen for a given area, retaining its biological diversity to the greatest degree possible is a sound strategy for ensuring its long-term survival and productivity. As plant ecologist Jerry Franklin of the University of Washington writes:

> We could never hope to adequately protect biological diversity solely through preservation.... The productivity of our land, the diversity of our plant and animal gene pool, and the overall integrity of our forest and stream ecosystems must be protected on [commodity-producing] landscapes as well as in preserves. Protection of diversity must be incorporated into everything we do every day on *every* acre...[91]

Toward Ecological Integrity

43

Since the days when coal miners carried canaries underground to warn of the presence of methane gas, living things have been used as indicators of broader environmental threats. One of the most celebrated cases is scientists' use of the northern spotted owl of the U.S. Pacific Northwest as an "indicator species" for the health of the region's ancient forest ecosystems. Its habitat decimated and its populations continuing to fall, the owl has also become a symbol of the nation's tattered biological systems.

In theory, the rare bird and its habitat are protected under a number of laws, including the Endangered Species Act, which prohibits destroying the habitat of any species determined to be in danger of extinction. But the government sought to skirt the laws protecting the owl. As of this writing, the so-called "God squad," convened to decide whether to consign the owl to oblivion because of the economic costs of saving the last old-growth stands, had yet to render its decision, and political impetus to weaken the Endangered Species Act continued to grow.[92]

The spotted owl—like the Philippine eagle, the Siberian crane, and the Honduran Emerald hummingbird—is a canary in a global coal mine, delivering a message that something is going wrong in the environment that supports us all. For the Pacific Northwest, it signals the imminent end of an era of natural resource mining, as ancient forests and the jobs that depend on their liquidation continue to dwindle. And like other endangered species conflicts developing in the United States, the decline of the owl is only a symptom of a broader mismanagement of natural resources that will end soon enough, whether we act to stop it or the resources being misused simply give out. But Americans and others who have benefited from the liquidation of biological assets seem reluctant to face up to the fundamental changes required to save endangered species, not to mention the global environment. Unwilling to hear the message that our all-consuming economy cannot be sustained, they angrily blame the messenger.

Perhaps the most essential step toward ecological integrity is recognizing the principle that biological diversity is valuable wherever it is found, and its loss anywhere is reason for concern. Diversity is an important, if often overlooked, indicator of environmental health, one that we ignore at our peril. Stemming the erosion of life on earth will require not only protecting nature reserves and indigenous homelands, but integrating the protection of diversity into the "mainstream" patterns of production, consumption, and waste disposal as well. In short, the conservation of biodiversity should guide all economic development.

What needs to be done to prevent biological decline is not always obvious. The basic challenges are to protect wilderness, manage other areas (such as timberlands, farms, and cities) in ways that maintain or restore diversity, and ensure that the costs and benefits of conserving the earth's biological wealth are divided equitably. How to accomplish these goals, however, will vary greatly from place to place. No single set of policy prescriptions will work to reconcile diverse societies and varied ecosystems to each other. No monolithic biodiversity commission can be charged with caretaking life for the world. The key will be an array of local approaches, informed by experiences in other places and supported by policies at higher levels.

The universal requirements for conservation, ironically, are those that seemingly have little to do with biological diversity. To intelligently limit the amount of the planet we dominate, and to tolerate diversity more in the places we do dominate, will entail tackling two of the most intractable and fundamental forces in the modern world: galloping per-capita consumption and rapid population growth. No conservation strategy, however ingenious, can get around the fact that the more resources one species consumes, the fewer are available for all the rest.

With so many forces working simultaneously to unravel the fabric of biodiversity, a setting of priorities is necessary. Efforts to expand strictly protected areas or accommodate diversity in other areas will be most effective if they are based on solid biological information. A technique known as gap analysis, for example, is being used in the United States to inventory and focus protection efforts on those types of natural communities that are least represented in protected areas or have special biological values. Even within biologically rich regions, some areas deserve

> "To intelligently limit the amount of the planet we dominate will entail tackling two of the most intractable and fundamental forces in the modern world: galloping per-capita consumption and rapid population growth."

more protection than others. At a landmark meeting in Manaus, Brazil in 1990, more than 90 scientists from a variety of disciplines pooled their knowledge of life in the Amazon basin to produce a detailed map of biological priorities for conservation in the region. If put to use by decision makers, this information could greatly increase the effectiveness of conservation efforts in the world's greatest tropical forest region.[93]

Historic international negotiations that could raise biodiversity higher on the international agenda and set priorities for its protection unfortunately seem mired in political avoidance of basic reform. A global convention on forests, which was to be signed in time for the June 1992 U. N. Conference on Environment and Development (UNCED), has been downgraded to a "statement of principles"—and the governments involved apparently cannot even agree to that. As noted previously, the current negotiations toward a biodiversity convention appear to be at an impasse as industrial nations, led by the United States, resist footing the bill for conservation or changing the rules under which their genetic resource industries have profited. Developing nations, meanwhile, emphasize the importance of national sovereignty over genetic resources without recognizing the need to support local people's efforts to conserve and use genetic diversity.[94]

Progress at the international level is halting at best, but a number of national governments have taken promising steps to foster conservation by local people. In 1991, extensive Native American homelands were given official recognition by the governments of Canada, Brazil and Venezuela. In July, Venezuela recognized eight million hectares of forest as the permanent homeland of the beleaguered Yanomami people, and the Brazilian government followed suit four months later by recognizing a slightly larger area for the Yanomami living across the border. Together the adjacent reserves encompass an area more than three times the size of Costa Rica. If the governments follow up on the legal recognition of these areas by helping the Yanomami keep Brazilian *garimpeiros* (gold miners) and other intruders off their lands, the extinction of yet another Amazon tribe, and the diversity it has stewarded, may be prevented.[95]

Legal recognition of lands, resources, and intellectual rights will not ensure the survival of traditional ways, however. A deeper respect is also necessary. Indigenous knowledge often fades when the younger members of a culture fail to take interest in the ways of aged medicine men, midwives and other wisdom keepers. In Belize, scientists seeking to tap elderly healers' expertise in "bush medicine" recognize that these elders often value veneration over material gain. Under an agreement between the U.S. National Cancer Institute, New York Botanical Garden (NYBG), and the Ix Chel Tropical Research Center in San Ignacio, Belize, if any new drugs are developed out of the collaborative effort, a portion of the profits will be returned to Belize. More immediately, NYBG ethnobotanist Michael Balick and colleagues hold ceremonies to pay homage to the healers and to raise their status in the community. The researchers also give lectures for local youths about the importance of their seniors' knowledge. The packed audiences that the lectures usually draw provide hope that the healers' knowledge of rain forest plants will be passed on to a younger generation.[96]

Traditional knowledge, of course, does not exist in a vacuum. New pressures, such as encroachment by outsiders, growing population, and expanding markets for local products, often mean that the observance of old taboos or traditions is no longer enough to protect the environment. Joint management of natural resources by governments and communities is one promising way to ensure that local people benefit from the insights of both traditional knowledge and Western science. The Brazilian government, for example, recently announced a plan to use satellite imagery, and a computerized geographic information system being completed this spring, to help the Yanomami and other tribes detect *garimpeiros* and drug smugglers on their lands.[97]

It remains doubtful, however, whether Brazil will tackle the basic social ills that drive *garimpeiros* to the frontier in the first place, or work to stop the violence suffered by rural activists at the hands of the hired gunmen of powerful landowners. In Brazil, as in many parts of the world, to speak up for the environment is literally to put one's life in danger. According to the human rights group Americas Watch, more than 1500 rural people (the most famous being rubber tapper Chico Mendes) have been killed in land disputes in the Brazilian interior over the past twenty-five years. In February 1991, 14 members of the Philippines' largest environmental group, the Haribon Foundation, were arrested after exposing government and military involvement in illegal logging on the island of Palawan. This past February, in the Malaysian state of Sarawak, police arrested Anderson Mutang Urud, director of the Sarawak Indigenous People's Alliance, without charge, apparently in order to intimidate members of the Penan tribe into dismantling their blockades of logging roads through their rain forest homeland.[98]

As these examples make clear, diversity conservation and community-based management of natural resources depend on human rights and basic political freedoms, which are sorely lacking in many countries, including many that call themselves democracies. As long as governments have limited accountability to their citizens, "parastatal" logging, mining, and fishing companies, often run by close friends or relatives of top politicians, will continue to have relatively free rein over public resources. Half the logging concessions in Sarawak, for example, are owned by family and friends of the current and former chief ministers of the state. And as long as access to environmental information and the freedom to put that information to use are restricted, the ability of citizens or non-governmental organizations (NGOs) to monitor or influence environmental abuses will be limited.[99]

An overhaul of humanity's relationship to nature will only be possible with an overhaul of the relationships among people themselves. Because developing nations generally lack the money and the technology to protect their living heritage, and because widespread poverty discourages a fifth of the planet's people from worrying about anything more than immediate needs, financial transfers from North to South are an essential component of conservation. And since much of the devastation of Southern economies and ecosystems is driven by Northern consumption of their resources, inequities of trade, debt and perhaps most important of all, overconsumption by wealthy nations and individuals, need to be tackled.

In both industrial and developing nations, more funding will be required for nature and genetic conservation. While biodiversity—the most basic of all resources underlying human economies—is a matter of national security, relatively little is spent to secure its future. The $340 million spent by the U. S. Fish and Wildlife Service to administer the Endangered Species Act over the past 17 years, for example, totaled less than the Sandia National Laboratory spent on nuclear weapons research in 1991 alone. Estimates of the funding needed to implement a global biodiversity convention vary greatly and range as high as $50 billion annually. But even the most inflated estimates are dwarfed by many of the destructive luxuries the world lavishes money on. World military expenditures in 1991 were $980 billion.[100]

In this context, the 1990 birth of the Global Environment Facility (GEF), an international fund for environmental protection that is expected to spend perhaps a third of its $1.3 billion budget on biodiversity conservation projects, was a welcome event. But the newborn GEF already appears to be taking on certain unwelcome characteristics of its primary administrator, the World Bank. (The U. N. Development Programme and U. N. Environment Programme also share in GEF's management.) Many of the proposed GEF projects are linked to traditional large-scale World Bank development projects, whose ecological impacts may negate any benefits realized through the add-on GEF programs. Outside access to information on the projects and popular input to their design have been limited, although this may be changing due to growing criticisms of the program.[101]

A more promising model for international conservation financing may be the recently-established Foundation for the Philippine Environment, a $25 million endowment for biological conservation, community-based management of natural resources, and institution building set up by the Philippine Department of Environment and Natural Resources, the U. S. Agency for International Development and a variety of NGOs in both nations. Unlike many governmental agreements, which tend to allow input of citizens or private groups as an afterthought at best, this fund allowed for active involvement of NGOs from the beginning. The majority of the Foundation's governing board (which will decide where money will be spent) is to be elected by members of Philippine NGOs, and these groups are expected to eventually take over the fund's administration, thus promising a greater accountability to local people's concerns.[102]

It is clear that protecting biodiversity will have a cost, as all things of value do; and when the value is incalculable—as in this realm it is—the cost may be steep. It is also clear that there will be immense economic benefits in its protection, and human suffering in its continued loss. The benefits will include money saved when destructive subsidies (for logging, for example) are halted, and when other destructive expenditures (for heavy pesticide use in agriculture, for example) are redirected. The benefits also include the array of existing and new products (medicines, foods, and the like) made available by conserving and studying natural ecosystems rather than blindly exploiting them.

But to try to balance the costs and benefits would be to miss a larger point: life on earth transcends economics, even though economic insights can help choose effective means of halting biological losses. No price can be assigned to the ability of the atmosphere, forests, and oceans together to maintain a life-giving climate; no value can be assigned to a species that has endured for millions of years. Moreover, a viable relationship with the myriad parts and processes of the biosphere lies not so much in any economic sacrifice *for* them as in a recognition of our dependence *on* them, and a willingness to let this insight guide all our activities.

Making the conservation of diversity a goal in everything we do would indeed be a fundamental shift. But anything less would be an abdication of our obligation to pass on to future generations a world of undiminished options, and of our moral responsibility as travelers on the only planet known to support life.

Notes

1. Species numbers from Nigel E. Stork, "Insect Diversity: Facts, Fiction and Speculation," *Biological Journal of the Linnean Society*, Vol. 35, 1988, and from E.O. Wilson, "The Current State of Biological Diversity," in E.O. Wilson and Frances M. Peter, eds., *Biodiversity* (Washington, D.C.: National Academy Press, 1988); soil from John Lancaster, "Protecting a Wealth of Species," *Washington Post*, May 24, 1991.

2. Edward O. Wilson, *The Diversity of Life* (Cambridge, Mass: Harvard University Press, forthcoming), based on assumed total of 10 million tropical rain forest species. For other recent scientific estimates, see Walter V. Reid, "How Many Species Will There Be?" in T. Whitmore and J. Sayer, eds., *Tropical Deforestation and Species Extinction* (London: Chapman & Hall, 1992), which predicts that deforestation will commit 0.1-0.5 percent of tropical forest species (or 10,000-50,000 species out of 10 million) to extinction per year, and Jared Diamond, "Playing Dice with Megadeath," *Discover*, April 1990, which assumes that current trends will result in the loss of 90 percent of tropical forests and half their 30 million species over the next century, or 150,000 species per year.

3. World Resources Institute (WRI) et al., *Global Biodiversity Strategy: Guidelines for Action to Save, Study, and Use Earth's Biotic Wealth Sustainably and Equitably* (Washington, D.C.: 1992); Larry Harris, Department of Wildlife and Range Sciences, University of Florida, Gainesville, Fla., private communication, July 7, 1991.

4. Background rates from Walter V. Reid and Kenton R. Miller, *Keeping Options Alive: The Scientific Basis for Conserving Biodiversity* (Washington, D.C.: WRI, 1989).

5. Aldo Leopold, *A Sand County Almanac, With Essays From Round River* (New York: Sierra Club/Ballantine Books, 1966).

6. Species from Reid and Miller, *Keeping Options Alive*; loss of nearly half of tropical rain forests from Norman Myers, *Deforestation Rates in Tropical Forests and Their Climatic Implications* (London: Friends of the Earth, 1989); 17 million hectares from William Booth, "Tropical Forests Disappearing at Faster Rate," *Washington Post*, September 9, 1991; Ismail Serageldin, *Saving Africa's Rainforests* (Washington, D.C.: World Bank, 1991).

7. James Brooke, "Amazon Forest Loss is Sharply Cut in Brazil," *New York Times*, March 26, 1991; Atlantic forests from Mark Collins, ed., *The Last Rain Forests* (London: Mitchell Beazley, 1990); John R. McNeill, "Deforestation in the Araucaria Zone of Southern Brazil, 1900-1983," in John F. Richards and Richard P. Tucker, eds., *World Deforestation in the Twentieth Century* (Durham, N.C.: Duke University, 1988).

8. Prairies from *Final Consensus Report of the Keystone Policy Dialogue on Biological Diversity on Federal Lands* (Keystone, Colo.: Keystone Center, 1991); cedars from Reid and Miller, *Keeping Options Alive*; Sandra Postel and John C. Ryan, "Reforming Forestry," in Lester R. Brown et al., *State of the World 1991* (New York: W.W. Norton & Co., 1991); Spencer B. Beebe, "Conservation in Temperate and Tropical Rain Forests: The Search for an Ecosystem Approach to Sustainability," paper presented at the 56th North American Wildlife and Natural Resources Conference, Edmonton, Alberta, Canada, March 25-29, 1991; Keith Moore, *Coastal Watersheds: An Inventory of Watersheds in the Coastal Temperate Forests of*

British Columbia (Vancouver: Earthlife Canada Foundation and Ecotrust/Conservation International, 1991).

9. Gianfranco Bologna, Fondo Mondiale per la Natura, Rome, private communication, March 16, 1992; Joanna Gould, Maruia Society, Nelson, New Zealand, private communication, March 17, 1992; Thomas E. Dahl, *Wetlands Losses in the United States 1780's to 1980's* (Washington, D.C.: U.S. Department of the Interior, Fish and Wildlife Service, 1990); "Sustaining Wetlands: International Challenge for the 90s," final report of the Sustaining Wetlands Forum, Ottawa, Ont., Canada, April 1990; prairie wetlands from Peter Lee, "Viewpoint," Environment Alberta, *Environment Views*, September 1990.

10. One-fourth figure from Joanna C. Ellison and David R. Stoddart, "Mangrove Ecosystem Collapse During Predicted Sea-Level Rise: Holocene Analogues and Implications," *Journal of Coastal Research*, Winter 1991; heavy losses from Don Hinrichsen, *Our Common Seas: Coasts in Crisis* (London: Earthscan, 1990); Ecuador also from Monica Herzig Zürcher and Alejandro Toledo Ocampo, "The Nineties: Another 'Lost Decade' for Latin American Wetlands?" *IWRB News* (International Waterfowl and Wetlands Research Bureau), January 1991; *Biodiversity Action Plan for Indonesia* (draft) (Ciloto, W. Java, Indonesia: National Planning Agency et al., 1991).

11. Boyce Thorne-Miller and John Catena, *The Living Ocean: Understanding and Protecting Marine Biodiversity* (Washington, D.C.: Island Press, 1991); Pamela Hallock-Muller, "Coastal Pollution and Coral Communities," *Underwater Naturalist*, Vol. 19, No. 1, 1990.

12. Susan M. Wells, ed., *Coral Reefs of the World* (Gland, Switzerland: International Union for the Conservation of Nature (IUCN) and United Nations Environment Programme (UNEP), 1988); Jon Miller, "Troubled Waters," *Far Eastern Economic Review*, March 15, 1990; Philippines diversity from Les Kaufman, "Marine Biodiversity: The Sleeping Dragon," *Conservation Biology*, December 1988.

13. Bleaching dates from Ernest H. Williams Jr. and Lucy Bunkley-Williams, "Coral Reef Bleaching Alert," *Nature*, July 19, 1990; Raymond L. Hayes and Thomas J. Goreau, "The Tropical Coral Reef Ecosystem as a Harbinger of Global Warming," paper presented at the 2nd International Conference on Global Warming, Chicago, April 8-11, 1991.

14. "No Money for Hawaiian Snails," *Oryx*, January 1991. The sources for Table 2 are: amphibians from David B. Wake, "Declining Amphibian Populations," *Science*, August 23, 1991, from David Towns and Ian Atkinson, "New Zealand's Restoration Ecology," *New Scientist*, April 20, 1991, and from Radhakrishna Rao, "India: Bullfrog Extinction," *Third World Week*, November 23, 1990; birds from N. J. Collar and P. Andrew, *Birds to Watch: The ICBP Checklist of Threatened Birds* (Cambridge, U.K.: International Council for Bird Preservation, 1988), and from Barry R. Noon and Kimberly Young, "Evidence of Continuing Worldwide Declines in Bird Populations: Insights from an International Conference in New Zealand," *Conservation Biology*, June 1991; fish from Jack E. Williams et al., "Fishes of North America Endangered, Threatened, or of Special Concern: 1989," *Fisheries*, November/December 1989, from "Fishing and Pollution Imperil Coastal Fish, Several Studies Find," *New York Times*, July 16, 1991, and from Leslie S. Kaufman et al., "A

52

Decade of Ecological Change in Lake Victoria," New England Aquarium, Boston, Mass., unpublished; invertebrates from Wilson, *The Diversity of Life*, from Organisation for Economic Co-operation and Development (OECD), *State of the Environment 1991* (Paris: 1991), and from WRI et al., *Global Biodiversity Strategy*; mammals from W. L. Hare, ed., *Ecologically Sustainable Development* (Fitzroy, Australia: Australian Conservation Foundation et al., 1990), and from OECD, *State of the Environment 1991*; carnivores from "Saving the Wild Cats: A Manifesto on Cat Conservation," *Tiger Paper*, January/March 1990, and from Jane E. Brody, "Boom in Poaching Threatens Bears Worldwide," *New York Times*, May 1, 1990; threatened primates from IUCN, *1990 IUCN Red List of Threatened Animals* (Gland, Switzerland, and Cambridge, U.K.: 1990) and from Russell A. Mittermeier, "Primate Diversity and the Tropical Forest," in Wilson and Peter, eds., *Biodiversity*; reptiles from Species Survival Commission, Tortoise and Freshwater Turtle Specialist Group, *Tortoises and Freshwater Turtles: An Action Plan for their Conservation*, 2nd ed. (Gland, Switzerland: IUCN, 1991), and, for sea turtles, from IUCN, *1990 IUCN Red List of Threatened Animals*.

15. G. Carleton Ray, "Coastal-Zone Biodiversity Patterns," *Bioscience*, July/August 1991.

16. Soviet Union from Carl Safina and Ken Hinman, "Stemming the Tide: Conservation of Coastal Fish Habitat in the United States," Summary of a National Symposium on Coastal Fish Habitat Conservation, Baltimore, Md., March 7-9, 1991; Malaysia from Jared M. Diamond, "The Present, Past and Future of Human-Caused Extinctions," *Philosophical Transactions of the Royal Society of London*, Vol. B 325, 1989.

17. UNEP, *Environmental Data Report*; Chris McIvor, "Namibia Fights to Save its Fish," *New African*, May 1991.

18. "Cutting Down the 'Walls of Death'," *Asiaweek*, December 20, 1991; Julian Baum, "Refusing the Bait," *Far Eastern Economic Review*, January 23, 1992; shrimp trawls from Eugene C. Bricklemyer, Jr., et al., "Discarded Catches in U.S. Commercial Fisheries," in William J. Chandler, ed., *Audubon Wildlife Report 1989/1990* (San Diego: Academic Press, 1989); yellowtail flounder trawls from Steven A. Murawski, "Can We Manage Our Multispecies Fisheries?" *Fisheries*, September/October 1991.

19. Wake, "Declining Amphibian Populations"; Marcia Barinaga, "Where Have All the Froggies Gone?" *Science*, March 2, 1990; Rao, "India: Bullfrog Extinction."

20. Reid and Miller, *Keeping Options Alive*; Peter H. Raven, "Biology in an Age of Extinction: What Is Our Responsibility?" Plenary Address, Fourth International Congress of Systematic and Evolutionary Biology, College Park, Md., July 1-4, 1990.

21. Southern Africa (which includes Botswana, Lesotho, Namibia, South Africa, and Swaziland) from Jeffrey A. McNeely et al., *Conserving the World's Biological Diversity* (Gland, Switzerland, and Washington, D.C.: IUCN et al., 1990); "Hundreds of Plant Species Threatened by Fungal Plague," Hong Kong AFP, March 24, 1991, transcribed in Foreign Broadcast Information Service, *Environmental Issues*, May 7, 1991; Philip Shabecoff, "Plant Lovers' Ambitious Goal Is No More Extinctions," *New York Times*, November 13, 1990.

22. Reid and Miller, *Keeping Options Alive*.

23. Cary Fowler and Pat Mooney, *Shattering: Food, Politics, and the Loss of Genetic Diversity* (Tucson: University of Arizona Press, 1990); Indonesian rice from *Biodiversity Action Plan for Indonesia*, and from WRI et al., *Global Biodiversity Strategy*.

24. Reid and Miller, *Keeping Options Alive*; Victoria Griffith, "Diseases Put Brazil's Oranges at Risk," *Financial Times*, June 14, 1991; Stephen B. Brush, "Farmer Conservation of New World Crops: The Case of Andean Potatoes," *Diversity*, Vol. 7, Nos. 1 and 2, 1991.

25. Stephen J. O'Brien et al., "The Cheetah in Genetic Peril," *Scientific American*, May 1986; Willa Nehlsen et al., "Pacific Salmon at the Crossroads: Stocks at Risk from California, Oregon, Idaho, and Washington," *Fisheries*, March/April 1991; William K. Stevens, "Hatched and Wild Fish: Clash of Cultures," *New York Times*, July 23, 1991.

26. World surface area from Mark S. Hoffman, ed., *The World Almanac and Book of Facts 1989* (New York: Pharos Books, 1988).

27. IUCN, *1990 United Nations List of National Parks and Protected Areas* (Gland, Switzerland, and Cambridge, U.K.: 1990); Chile from George Ledec, "A Proposed Strategy for the World Bank to Promote Increased Conservation of Biological Diversity in Latin America and the Caribbean," unpublished paper, April 1989; M.I. Dyer and M.M. Holland, "The Biosphere-Reserve Concept: Needs for A Network Design," *Bioscience*, May 1991; John C. Ryan, "Belize's Reefs on the Rocks," *World Watch*, November/December 1991.

28. Parks from World Conservation Monitoring Center (WCMC), *Global Biodiversity 1992* (draft) (Cambridge, U.K.: forthcoming); Janis Alcorn and Augusta Molnar, "Deforestation and Forest-Human Relationships: What Can We Learn from India?" paper presented at meeting of the American Anthropological Association, New Orleans, La., November 1990; James Rush, *The Last Tree: Reclaiming the Environment in Tropical Asia* (New York: The Asia Society, 1991); Kuna from Mac Chapin, "Losing the Way of the Great Father," *New Scientist*, August 10, 1991; William Harp, "Ecology and Cosmology, Rain Forest Exploitation Among the Emberá-Chocó," paper presented at Humid Tropical Lowlands Conference: Development Strategies and Natural Resource Management, Panama City, Panama, June 17-21, 1991; Tukano from Jason W. Clay, *Indigenous People and Tropical Forests: Models of Land Use and Management from Latin America* (Cambridge, Mass.: Cultural Survival, 1988).

29. J. Michael McCloskey and Heather Spalding, "A Reconnaissance-Level Inventory of the Amount of Wilderness Remaining in the World," *Ambio*, Vol. 18, No. 4, 1989; Janis B. Alcorn, "Ethics, Economies, and Conservation," in Margery L. Oldfield and Janis B. Alcorn, eds., *Biodiversity: Culture, Conservation and Ecodevelopment* (Boulder, Colo.: Westview Press, 1991); Arturo Gómez-Pompa and Andrea Kaus, "Conservation by Traditional Cultures in the Tropics," in Vance Martin, ed., *For the Conservation of Earth* (Golden, Colo.: Fulcrum, 1988).

30. Richard Evans Schultes, "Ethnobotanical Conservation and Plant Diversity in the Northwest Amazon," *Diversity*, Vol. 7, Nos. 1 and 2, 1991; tribes from Eugene Linden, "Lost

Tribes, Lost Knowledge," *Time*, September 23, 1991; Charles R. Clement, "Amazonian Fruits: Neglected, Threatened and Potentially Rich Resources Require Urgent Attention," *Diversity*, Vol. 7, Nos. 1 and 2, 1991.

54

31. John G. Robinson and Kent H. Redford, eds., *Neotropical Wildlife Use and Conservation* (Chicago: University of Chicago Press, 1991); Nigel Hicks, "Saving China's Last Tropical Rainforests," *Asiaweek*, July 12, 1991; Christopher A. Myers and Aniruddh D. Patel, "Saving Asia's Wildlife," *World Monitor*, January 1991.

32. Organization of American States (OAS) and National Park Service (NPS), U.S. Department of Interior, *Inventory of Caribbean Marine and Coastal Protected Areas* (Washington, D.C.: 1988); Mark Brazil, "Where Eastern Eagles Dare," *New Scientist*, May 4, 1991.

33. Jim Fulton, Member of Canadian Parliament, letter to Lucien Bouchard, Minister of Environment, February 6, 1990; Czechoslovakia from Jim Thorsell, "The IUCN Register of Threatened Protected Areas of the World," paper presented to the 34th Working Session of IUCN Commission on National Parks and Protected Areas, Perth, Australia, November 26-27, 1990; *Biodiversity Action Plan for Indonesia*; Olga Sheean, "Fool's Gold in Ecuador," *WWF News*, January/February 1992; Europe from Patrick C. West and Steven R. Brechin, eds., *Resident Peoples and National Parks: Social Dilemmas and Strategies in International Conservation* (Tucson: University of Arizona Press, 1991); marine areas from OAS and NPS, *Inventory of Caribbean Marine and Coastal Protected Areas*, and from "Focus on: Marine Protected Areas," in WRI, *World Resources 1988-89* (New York: Basic Books, 1988).

34. Poland from Thorsell, "The IUCN Register of Threatened Protected Areas"; Yasmin D. Arquiza, "Toll on the Atoll," *Far Eastern Economic Review*, March 15, 1990.

35. Peter Bruce, "EC Intervenes on Spain's Dying Wetlands," *Financial Times*, July 18, 1991.

36. John A. Dixon and Paul B. Sherman, *Economics of Protected Areas: A New Look at Benefits and Costs* (Washington, D.C.: Island Press, 1990); Alcorn, "Ethics, Economies, and Conservation."

37. Michael Wells and Katrina Brandon, *People and Parks: Linking Protected Area Management with Local Communities* (Washington, D.C.: World Bank et al., 1991).

38. Ibid.; "Inupiat Eskimos, Gwich'in Indians Disagree About Oil," *Christian Science Monitor*, July 30, 1991.

39. Mike De Mott, "Peru: Wildlife Area Focus of Ecological Debate," *Latinamerica Press*, June 20, 1991.

40. Wells and Brandon, *People and Parks*.

41. Daniel B. Botkin, *Discordant Harmonies: A New Ecology for the Twenty-First Century* (New York: Oxford University Press, 1990); on ancient clearance of seemingly pristine

Amazonian jungle, see Anna Roosevelt, "The Historical Perspective on Resource Use in Tropical Latin America," in Center for Latin American Studies, University of Florida, *Economic Catalysts to Ecological Change* (Gainesville, Fla.: 1990); toxic wastes have been found in fish of the deep ocean, see Thorne-Miller and Catena, *The Living Ocean*.

42. Monte Hummel, *A Conservation Strategy for Large Carnivores in Canada* (Toronto: World Wildlife Fund Canada, 1990); J. M. Thiollay, "Area Requirements for the Conservation of Rain Forest Raptors and Game Birds in French Guiana," *Conservation Biology*, June 1989; C.M. Pannell, "The Role of Animals in Natural Regeneration and the Management of Equatorial Rain Forests for Conservation and Timber Production," *Commonwealth Forestry Review*, Vol. 68, No. 4, 1989; Elliott A. Norse, *Ancient Forests of the Pacific Northwest* (Washington, D.C.: Island Press, 1990); Carlos A. Peres, "Humboldt's Woolly Monkeys Decimated by Hunting in Amazonia," *Oryx*, April 1991; Reed F. Noss, "Sustainability and Wilderness," *Conservation Biology*, March 1991.

43. Reed F. Noss, "What Can Wilderness Do For Biodiversity?" in P. Reed, ed., *Preparing to Manage Wilderness in the 21st Century* (Asheville, N.C.: U.S. Department of Agriculture, Forest Service, 1990); Craig L. Shafer, *Nature Reserves: Island Theory and Conservation Practice* (Washington, D.C.: Smithsonian Institution, 1990).

44. Adrian Barnett and Aléxia Celeste da Cunha, "The Golden-Backed Uacari on the Upper Rio Negro, Brazil," *Oryx*, April 1991.

45. Alan B. Durning, *Poverty and the Environment: Reversing the Downward Spiral*, Worldwatch Paper 92 (Washington, D.C.: Worldwatch Institute, November 1989); Holly B. Brough, "A New Lay of the Land," *World Watch*, January/February 1991.

46. Paul R. Ehrlich and Edward O. Wilson, "Biodiversity Studies: Science and Policy," *Science*, August 16, 1991; Reed Noss, "A Native Ecosystems Act (Concept Paper)," *Wild Earth*, Spring 1991; Z. Naveh, "Some Remarks on Recent Developments in Landscape Ecology as a Transdisciplinary Ecological and Geographical Science," *Landscape Ecology*, Vol. 5, No. 2, 1991.

47. Sixty percent of indigenous homelands in the Brazilian Amazon now have legal title, according to Carlos Alberto Ricardo, Centro Ecumênico de Documentação e Informação, São Paulo, Brazil, private communication, February 25, 1992; Alan Thein Durning, "Native Americans Stand Their Ground," *World Watch*, November/December 1991; Mark Poffenberger, ed., *Keepers of the Forest: Land Management Alternatives in Southeast Asia* (West Hartford, Conn.: Kumarian Press, 1990).

48. Sue Armstrong, "The People Who Want Their Parks Back," *New Scientist*, July 6, 1991; Alcorn, "Ethics, Economies, and Conservation."

49. Gregor Hodgson, "Drugs from the Sea," *Far Eastern Economic Review*, April 11, 1991; Philippe Rasoanaivo, "Rain Forests of Madagascar: Sources of Industrial and Medicinal Plants," *Ambio*, December 1990; Curt Suplee, "Medicine: Going to Bat Against Heart Attacks," *Washington Post*, July 15, 1991; Patricia Wright, Duke University, speech to

Smithsonian Resident Associates Program, Washington, D.C., January 28, 1991.

50. Christopher Joyce, "Prospectors for Tropical Medicines," *New Scientist*, October 19, 1991; William Booth, "U.S. Drug Firm Signs Up to Farm Tropical Forests," *Washington Post*, September 21, 1991.

51. Elaine Elisabetsky, "Sociopolitical, Economical and Ethical Issues in Medicinal Plant Research," *Journal of Ethnopharmacology*, No. 32, 1991; Amazonian figure from WRI et al., *Global Biodiversity Strategy*; rattan from Jenne H. De Beer and Melanie J. McDermott, *The Economic Value of Non-timber Forest Products in Southeast Asia* (Amsterdam: Netherlands Committee for IUCN, 1989).

52. David Western, "Conservation Without Parks: Wildlife in the Rural Landscape," in David Western and Mary Pearl, eds., *Conservation for the Twenty-First Century* (New York: Oxford University Press, 1989); Fikret Berkes, ed., *Common Property Resources: Ecology and Community-Based Sustainable Development* (London: Belhaven Press, 1989); De Beer and McDermott, *The Economic Value of Non-timber Forest Products*; John Kurien, *Ruining the Commons and Responses of the Commoners: Coastal Overfishing and Fishermen's Actions in Kerala State, India* (Geneva: UN Research Institute for Social Development, 1991).

53. De Beer and McDermott, *The Economic Value of Non-timber Forest Products*; fish from Oliver T. Coomes, "Rain Forest Extraction, Agroforestry, and Biodiversity Loss: An Environmental History from the Northeastern Peruvian Amazon," paper presented to the 16th International Congress of the Latin American Studies Association (hereinafter, LASA Congress), Washington, D.C., April 6, 1991; Rodolfo Vasquez and Alwyn H. Gentry, "Use and Misuse of Forest-harvested Fruits in the Iquitos Area," *Conservation Biology*, December 1989; 4 cents figure from Orna Feldman, "Rain Forest Chic," *New Republic*, June 25, 1990; 2-3 percent figure from Cultural Survival, "Cultural Survival Enterprises Direct Assistance Projects FY1989-1990," Cambridge, Mass., no date; market control from Stephan Schwartzman, "Marketing of Extractive Products in the Brazilian Amazon," Environmental Defense Fund, Washington, D.C., October 29, 1990.

54. John O. Browder, "Social and Economic Constraints on the Development of Market-Oriented Extractive Reserves in Amazon Rain Forests," *Annals of Economic Botany*, forthcoming; Americas Watch, *Rural Violence in Brazil* (Washington, D.C.: Human Rights Watch, 1991); goal of one fourth of Brazilian Amazon from Luis Fernando Allegretti, Institute for Amazon Studies, Curitiba, Brazil, speech to LASA Congress, April 4, 1991; Guatemala from Conrad Reining, "Non-Timber Forest Products and the Peten, Guatemala: Why Extractive Reserves Are Critical for Both Conservation and Development," Conservation International, Washington, D.C., unpublished paper, March 1991; Peru from Coomes, "Rain Forest Extraction."

55. Import tax from Environmental Defense Fund, "Rubber Tappers Demonstrate in Brasilia," press release, Washington, D.C., April 2, 1991; plantations from Nigel J.H. Smith and Richard Evans Schultes, "Deforestation and Shrinking Crop Gene-Pools in Amazonia," *Environmental Conservation*, Autumn 1990.

56. Douglas Daly, "Extractive Reserves: A Great New Hope," *Garden*, November/December 1990; Charles M. Peters, "Plenty of Fruit But No Free Lunch," *Garden*, November/December 1990.

57. Peters, "Plenty of Fruit"; Henrik Borgtoft Pedersen and Henrik Balslev, "Economic Botany of Ecuadorean Palms," paper presented at conference arranged by Conservation International and Asociación Nacional para la Conservación de la Naturaleza on "The Sustainable Harvest and Marketing of Rain Forest Products," Panama City, Panama, June 20-21, 1991.

58. Kurien, *Ruining the Commons*; G. Carleton Ray, "Ecological Diversity in Coastal Zones and Oceans," in Wilson and Peters, eds., *Biodiversity*; Robinson and Redford, eds., *Neotropical Wildlife Use and Conservation*.

59. Charles M. Peters et al., "Oligarchic Forests of Economic Plants in Amazonia: Utilization and Conservation of an Important Tropical Resource," *Conservation Biology*, December 1989; Christine Padoch, Institute of Economic Botany, New York Botanical Garden, speech to American Association for the Advancement of Science (AAAS) Annual Meeting, Washington, D.C., February 18, 1991.

60. Agnes Kiss, ed., *Living With Wildlife: Wildlife Resource Management with Local Participation in Africa* (Washington, D.C.: The World Bank, 1990).

61. Ibid.; Oliver Kanene, "Poacher's Pay-Off," *Panoscope*, January 1991; Zimbabwe figure from Mary-Lu Cole, "A Farm on the Wild Side," *New Scientist*, September 8, 1990.

62. Forest ownership figure from Robert Repetto, "Deforestation in the Tropics," *Scientific American*, April 1990; Mark Poffenberger, "Facilitating Change in Forestry Bureaucracies," in Poffenberger, ed., *Keepers of the Forest*; Conner Bailey and Charles Zerner, "Role of Traditional Fisheries Resource Management Systems for Sustainable Resource Utilization," presented at Forum Perikanan Dalam Pembangunan Jangka Panjang Tahap II: Tantangan dan Peluang, Sukabumi, West Java, Indonesia, June 18-21, 1991; Dale Lewis et al., "Wildlife Conservation Outside Protected Areas—Lessons from an Experiment in Zambia," *Conservation Biology*, June 1990.

63. Figure of 97 percent from Arthur Mitchell et al., "Community Participation for Conservation Area Management in the Cyclops Mountains, Irian Jaya, Indonesia," in Poffenberger, ed., *Keepers of the Forest*; George Marshall, "The Political Economy of Logging: The Barnett Inquiry into Corruption in the Papua New Guinea Timber Industry," *The Ecologist*, September/October 1990; "Putting Teeth in Logging Laws," *Asiaweek*, August 30, 1991.

64. Postel and Ryan, "Reforming Forestry"; A.J. Hansen et al., "Conserving Biodiversity in Managed Forests: Lessons from Natural Forests," *Bioscience*, June 1991; Ken Lertzman, "What's New About *New Forestry*? Replacing Arbocentrism in Forest Management," *Forest Planning Canada*, May-June 1990.

65. Pannell, "The Role of Animals in Natural Regeneration"; Postel and Ryan, "Reforming Forestry"; Marguerite Holloway, "Hol Chan: Marine Parks Benefit Commercial Fisheries," *Scientific American*, May 1991; A.T. White, "The Effect of Community-Managed Marine Reserves in the Philippines on Their Associated Coral Reef Fish Populations," *Asian Fisheries Science*, Vol. 2, 1988; Wendell Berry, "Preserving Wildness," *Wilderness*, Spring 1987.

66. Genetic vulnerability from Gary Paul Nabhan, *Enduring Seeds: Native American Agriculture and Wild Plant Conservation* (San Francisco, Calif.: North Point Press, 1989), and from Fowler and Mooney, *Shattering*.

67. Earl Young, Belize Fisheries Department, Coastal Zone Management Unit, Belize City, private communication, July 2, 1991; Ann Y. Robinson, "Conservation Compliance and Wildlife," *Journal of Soil and Water Conservation*, January/February 1989; David Baldock, *Agriculture and Habitat Loss in Europe* (Gland, Switzerland: World Wide Fund for Nature International, 1990); B.H. Green, "Agricultural Impact on the Rural Environment," *Journal of Applied Ecology*, Vol. 26, 1989; Robert M. May, "Avian Analyses," *Nature*, October 25, 1990.

68. Judith D. Soule and Jon K. Piper, *Farming in Nature's Image: An Ecological Approach to Agriculture* (Washington, D.C.: Island Press, 1992); David F. Bezdicek and David Granatstein, "Crop Rotation Efficiencies and Biological Diversity in Farming Systems," *American Journal of Alternative Agriculture*, Vol. 4, Nos. 3 and 4, 1989; Peter G. Kevan et al., "Insect Pollinators and Sustainable Agriculture," *American Journal of Alternative Agriculture*, Vol. 5, No. 1, 1990; D. L. Hawksworth, *The Biodiversity of Microorganisms and Invertebrates: Its Role in Sustainable Agriculture* (Wallingford, U.K.: CAB International, 1991).

69. Baldock, *Agriculture and Habitat Loss in Europe*; Robinson, "Conservation Compliance and Wildlife"; Defenders of Wildlife, *In Defense of Wildlife: Preserving Communities and Corridors* (Washington, D.C.: 1989).

70. Subsidies from WRI et al., *Global Biodiversity Strategy*; Brough, "A New Lay of the Land."

71. Keystone Center, *Oslo Plenary Session Final Consensus Report: Global Initiative for the Security and Sustainable Use of Plant Genetic Resources* (Oslo: 1991); Damien Lewis, "The Gene Hunters," *Geographical Magazine*, January 1991; "Delegates Pessimistic About Outlook for Finishing Work on Biodiversity Treaty," *International Environment Reporter*, February 26, 1992.

72. Nabhan, *Enduring Seeds*; Miguel A. Altieri, "Traditional Farming in Latin America," *The Ecologist*, March/April 1991; Oldfield and Alcorn, *Biodiversity*.

73. Altieri, "Traditional Farming"; Robert Chambers et al., eds., *Farmer First: Farmer Innovation and Agricultural Research* (London: Intermediate Technology, 1989); Patricia Allen and Debra Van Dusen, eds., *Global Perspectives on Agroecology and Sustainable Agricultural Systems*, Proceedings of the 6th International Scientific Conference of the

International Federation of Organic Agriculture Movements (Santa Cruz: University of California, 1989).

74. See the "Sustainable Urban Landscape" issue of *Earthword*, Fall 1991; Reed F. Noss, "Wildlife Corridors," in D. Smith, ed., *Ecology of Greenways* (Minneapolis: Univ. of Minnesota, forthcoming); Lowell W. Adams and Louise E. Dove, *Wildlife Reserves and Corridors in the Urban Environment: A Guide to Ecological Landscape Planning and Resource Conservation* (Columbia, Md.: National Institute for Urban Wildlife, 1989).

75. Kathleen Landauer and Mark Brazil, eds., *Tropical Home Gardens* (Tokyo: United Nations University, 1990); Kenneth A. Dahlberg, "Fields, Fisheries, Grasslands, Forests: Towards More Regenerative Systems," draft paper presented at AAAS annual meeting, Washington, D.C., February 17, 1991.

76. Anderson quoted in Nabhan, *Enduring Seeds*.

77. Boyce Rensenberger, "500,000 More Years of Human Antiquity," *Washington Post*, February 20, 1992; Andrew Hill et al., "Earliest *Homo*," *Nature*, February 20, 1992.

78. Carbon dioxide levels from U.S. Congress, Office of Technology Assessment, *Changing by Degrees: Steps to Reduce Greenhouse Gases* (Washington, D.C.: U.S. Government Printing Office, 1991); Intergovernmental Panel on Climate Change (IPCC), "Policymakers' Summary of the Potential Impacts of Climate Change," Report from Working Group II to IPCC, Geneva, 1990; Robert L. Peters and J. P. Myers, "Preserving Biodiversity in a Changing Climate," *Issues in Science and Technology*, Winter 1991-92.

79. Vegetation displacement and North American tree species movement from Jacqueline H. W. Karas, *Back from the Brink: Greenhouse Gas Targets for a Sustainable World* (London: Friends of the Earth, 1991); boreal forests from Allen M. Solomon, "Predicting Boreal and Temperate Forest Ecosystem Response to Climatic Change," abstract of lecture given to *Impacts of Climate Change on Ecosystems and Species* symposium, Amersfoort, Netherlands, December 2-6, 1991 and from R. T. Lester and J. P. Myers, "Global Warming, Climate Disruption, and Biological Diversity," in Chandler, ed., *Audubon Wildlife Report 1989/1990*.

80. Gary S. Hartshorn, "Possible Effects of Global Warming on the Biological Diversity in Tropical Forests," in Robert L. Peters and Thomas E. Lovejoy, eds., *Global Warming and Biological Diversity* (New Haven: Yale University Press, forthcoming).

81. Weeds and pests from Peters and Myers, "Preserving Biodiversity in a Changing Climate"; sea grasses from Joop Brouns, "The Impact of Climate Change on Seagrasses," abstract of lecture given to *Impacts of Climate Change on Ecosystems and Species* symposium, Amersfoort, Netherlands, December 2-6, 1991; disturbances from Lester and Myers, "Global Warming, Climate Disruption, and Biological Diversity." Sources for Table 6 include: coral bleaching from Rik Leemans, National Institute of Public Health and Environmental Protection, Bilthoven, Netherlands, "Impacts of Climate Change on Ecosystems and Species: Report from an International Symposium," unpublished manuscript, 1992, and from Walter V. Reid and Mark C. Trexler, *Drowning the National*

Heritage: Climate Change and U.S. Coastal Biodiversity (Washington, D.C.: WRI, 1991); ice melt effects from Adam Markham, *Can Nature Survive Global Warming?: Working Paper* (Gland, Switzerland: World Wide Fund for Nature, 1992); increased growth and metabolism from Robert C. Francis and Thomas H. Sibley, "Climate Change and Fisheries: What Are the Real Issues?" *Northwest Environmental Journal*, Fall/Winter 1991; animal migrations from Lester and Myers, "Global Warming, Climate Disruption, and Biological Diversity"; mangroves from Ellison and Stoddart, "Mangrove Ecosystem Collapse"; corals under rising seas from Markham, *Can Nature Survive Global Warming?*; shifting currents from Thorne-Miller and Catena, *The Living Ocean*; wave action and storms from Karas, *Back from the Brink*. Peruvian fisheries periodically collapse when abnormally warm temperatures cause upwelling currents to weaken: see H.V. Thurman, *The Essentials of Oceanography* (Columbus, Ohio: Merrill Publishing, 1987).

82. Freshwater from WRI et al., *Global Biodiversity Strategy*; Markham, *Can Nature Survive Global Warming?*

83. IPCC, *Climate Change: The IPCC Response Strategies* (Washington, D.C.: Island Press, 1991).

84. Ellison and Stoddart, "Mangrove Ecosystem Collapse."

85. Passenger pigeons from Sara L. Webb, "Potential Role of Passenger Pigeons and Other Vertebrates in Rapid Holocene Migrations of Nut Trees," *Quaternary Research*, Vol. 26:367, 1980; Dexter Hinckley and Geraldine Tierney, "Ecosystem Responses to Rapid Climate Change—Past and Future," unpublished paper, U.S. Environmental Protection Agency (EPA), Washington, D.C., August 1991.

86. Joel B. Smith and Dennis Tirpak, eds., *The Potential Effects of Global Climate Change on the United States: Report to Congress* (Washington, D.C.: EPA, 1989).

87. Halpin study in Thomas M. Smith et al., "Global Forests," in *Progress Reports on International Studies of Climate Change Impacts* (draft) (Washington, D.C.: EPA, 1990); Rik Leemans, National Institute of Public Health and Environmental Protection, Bilthoven, Netherlands, private communication, February 10, 1992; Markham, *Can Nature Survive Warming?*

88. Peters and Myers, "Preserving Biodiversity in a Changing Climate."

89. Research needs discussed in B.J. Huntley et al., "A Sustainable Biosphere: The Global Imperative," *Ecology International*, Vol. 20, 1991, and in *Conserving Biodiversity: A Research Agenda for Development Agencies* (Washington, D.C.: National Academy Press, 1992); Wendy E. Hudson, ed., *Landscape Linkages and Biodiversity* (Washington, D.C.: Island Press, 1991); Thorne-Miller and Catena, *The Living Ocean*.

90. W. L. Hare, ed., *Ecologically Sustainable Development* (Fitzroy, Australia: Australian Conservation Foundation et al., 1990).

91. Jerry Franklin, "Toward A New Forestry," *American Forests*, November/December 1989; David A. Perry and Jumanne Maghembe, "Ecosystem Concepts and Current Trends in Forest Management: Time for Reappraisal," *Forest Ecology and Management*, Vol. 26, 1989.

92. Kathryn A. Kohm, ed., *Balancing on the Brink of Extinction* (Washington, D.C: Island Press, 1991); Timothy Egan, "Hearings on the Spotted Owl Begin," *New York Times*, January 9, 1992.

93. Hudson, ed., *Landscape Linkages and Biodiversity*; Ghillean Prance, "Consensus for Conservation," *Nature*, May 31, 1990; Thomas E. Lovejoy et al., eds., *Priority Areas for Conservation in the Amazon* (Washington, D.C.: Smithsonian Press, forthcoming).

94. George M. Woodwell and Kilaparti Ramakrishna, "Forests, Scapegoats and Global Warming," *New York Times*, February 11, 1992; William K. Stevens, "Talks Seek to Prevent Huge Loss of Species," *New York Times*, March 3, 1992; Second European Network Meeting on Genetic Resources and Biotechnology, *Final Report* (Barcelona, Spain: Genetic Resources Action International, 1991).

95. "Something New in Canada's Frozen North," *The Economist*, January 4, 1992; Julia Preston, "Brazil Grants Land Rights to Indians," *Washington Post*, November 16, 1991; James Brooke, "Venezuela Befriends Tribe, but What's Venezuela?" *New York Times*, September 11, 1991; James Brooke, "Venezuela's Policy for Brazil's Gold Miners: Bullets," *New York Times*, February 16, 1992; James Brooke, "Brazil Steps Up Drive to Protect Indian Lands," *New York Times*, December 1, 1991.

96. Linden, "Lost Tribes, Lost Knowledge"; Michael J. Balick, "The Belize Ethnobotany Project: Discovering the Resources of the Tropical Rain Forest," *Fairchild Tropical Garden Bulletin*, April 1991; Michael J. Balick, Institute of Economic Botany, New York Botanical Garden, private communication, March 4, 1992.

97. Tom Kizzia, *The Wake of the Unseen Object: Among the Native Cultures of Bush Alaska* (New York: Henry Holt and Company, 1991); Christina Lamb, "Operation Amazon," *Financial Times*, January 17, 1992.

98. Americas Watch, *Rural Violence in Brazil*; "PNP Rapped for Arrest of Haribon Staff," *Philippine Daily Inquirer*, February 22, 1991; Western Canada Wilderness Committee, "Protest the Arrest of Anderson Mutang Urud," press release, Vancouver, February 10, 1992.

99. John C. Ryan, "Plywood Vs. People in Sarawak," *World Watch*, January/February 1991.

100. U.S. spending comparison from Brad Knickerbocker, "Extinctions 'Reduced to a Trickle'," *Christian Science Monitor*, December 24, 1991, and from Michael Bean, Environmental Defense Fund, Washington, D.C., private communication, March 18, 1992; Ruth Leger Sivard, *World Military and Social Expenditures 1991* (Washington, D.C.: World Priorities, 1991).

62

101. Budget figures from Michael Prowse, "Financing a Green Future in a Planet Without Borders," *Financial Times*, February 14, 1992; Global Environment Facility, Washington, D.C., "Monthly Operations Report," February 1992; criticisms from WRI et al., *Global Biodiversity Strategy* and from Jeffrey A. McNeely, "The Dangers of Too Much Money: The Global Environment Facility and Biodiversity," *Environmental Awareness*, October-December 1991.

102. Eugene Gonzales, The Caucus of Development NGO Networks, Manila, speech to Philippine Development Forum meeting, Washington, D.C., March 2, 1992.

JOHN C. RYAN is a Research Associate at the Worldwatch Institute, and coauthor of two of the Institute's *State of the World* reports. His research focuses on forests and biological diversity. He studied environmental history at Yale University and received a master's degree in history fom Stanford University.

THE WORLDWATCH PAPER SERIES

_____ 91. **Slowing Global Warming: A Worldwide Strategy** by Christopher Flavin.
_____ 92. **Poverty and the Environment: Reversing the Downward Spiral** by
Alan B. Durning.
_____ 93. **Water for Agriculture: Facing the Limits** by Sandra Postel.
_____ 94. **Clearing the Air: A Global Agenda** by Hilary F. French.
_____ 95. **Apartheid's Environmental Toll** by Alan B. Durning.
_____ 96. **Swords Into Plowshares: Converting to a Peace Economy** by
Michael Renner.
_____ 97. **The Global Politics of Abortion** by Jodi L. Jacobson.
_____ 98. **Alternatives to the Automobile: Transport for Livable Cities** by
Marcia D. Lowe.
_____ 99. **Green Revolutions: Environmental Reconstruction in Eastern
Europe and the Soviet Union** by Hilary F. French.
_____ 100. **Beyond the Petroleum Age: Designing a Solar Economy** by
Christopher Flavin and Nicholas Lenssen.
_____ 101. **Discarding the Throwaway Society** by John E. Young.
_____ 102. **Women's Reproductive Health: The Silent Emergency** by Jodi L. Jacobson.
_____ 103. **Taking Stock: Animal Farming and the Environment** by Alan B. Durning and
Holly B. Brough.
_____ 104. **Jobs in a Sustainable Economy** by Michael Renner.
_____ 105. **Shaping Cities: The Environmental and Human Dimensions** by Marcia D. Lowe.
_____ 106. **Nuclear Waste: The Problem That Won't Go Away** by Nicholas Lenssen.
_____ 107. **After the Earth Summit: The Future of Environmental Governance**
by Hilary F. French.
_____ 108. **Life Support: Conserving Biological Diversity** by John C. Ryan.

_____ **Total Copies**

☐ **Single Copy: $5.00**
☐ **Bulk Copies (any combination of titles)**
   ☐ 2–5: $4.00 each   ☐ 6–20: $3.00 each   ☐ 21 or more: $2.00 each

☐ **Membership in the Worldwatch Library: $25.00 (overseas airmail $40.00)**
The paperback edition of our 250-page "annual physical of the planet,"
State of the World 1991, plus all Worldwatch Papers released during
the calendar year.

☐ **Subscription to *World Watch* Magazine: $15.00 (overseas airmail $30.00)**
Stay abreast of global environmental trends and issues with our award-winning,
eminently readable bimonthly magazine.

No postage required on prepaid orders. Minimum $3 postage and handling
charge on unpaid orders.

Make check payable to Worldwatch Institute
1776 Massachusetts Avenue, N.W., Washington, D.C. 20036-1904 USA

Enclosed is my check for U.S. $_____

_____

name   **daytime phone #**

_____

address

_____

city   **state**   **zip/country**

five dollars

Worldwatch Institute
1776 Massachusetts Avenue, N.W.
Washington, D.C. 20036 USA